Takayuki Ito, Minjie Zhang, Valentin Robu, Shaheen Fatima, Tokuro Matsuo, and Hirofumi Yamaki (Eds.)

# Innovations in Agent-Based Complex Automated Negotiations

# Studies in Computational Intelligence, Volume 319

**Editor-in-Chief**
Prof. Janusz Kacprzyk
Systems Research Institute
Polish Academy of Sciences
ul. Newelska 6
01-447 Warsaw
Poland
*E-mail:* kacprzyk@ibspan.waw.pl

---

Further volumes of this series can be found on our homepage: springer.com

Vol. 295. Roger Lee (Ed.)
*Software Engineering, Artificial Intelligence, Networking and Parallel/Distributed Computing,* 2010
ISBN 978-3-642-13264-3

Vol. 296. Roger Lee (Ed.)
*Software Engineering Research, Management and Applications,* 2010
ISBN 978-3-642-13272-8

Vol. 297. Tania Tronco (Ed.)
*New Network Architectures,* 2010
ISBN 978-3-642-13246-9

Vol. 298. Adam Wierzbicki
*Trust and Fairness in Open, Distributed Systems,* 2010
ISBN 978-3-642-13450-0

Vol. 299. Vassil Sgurev, Mincho Hadjiski, and Janusz Kacprzyk (Eds.)
*Intelligent Systems: From Theory to Practice,* 2010
ISBN 978-3-642-13427-2

Vol. 300. Baoding Liu (Ed.)
*Uncertainty Theory,* 2010
ISBN 978-3-642-13958-1

Vol. 301. Giuliano Armano, Marco de Gemmis, Giovanni Semeraro, and Eloisa Vargiu (Eds.)
*Intelligent Information Access,* 2010
ISBN 978-3-642-13999-4

Vol. 302. Bijaya Ketan Panigrahi, Ajith Abraham, and Swagatam Das (Eds.)
*Computational Intelligence in Power Engineering,* 2010
ISBN 978-3-642-14012-9

Vol. 303. Joachim Diederich, Cengiz Gunay, and James M. Hogan
*Recruitment Learning,* 2010
ISBN 978-3-642-14027-3

Vol. 304. Anthony Finn and Lakhmi C. Jain (Eds.)
*Innovations in Defence Support Systems,* 2010
ISBN 978-3-642-14083-9

Vol. 305. Stefania Montani and Lakhmi C. Jain (Eds.)
*Successful Case-Based Reasoning Applications-1,* 2010
ISBN 978-3-642-14077-8

Vol. 306. Tru Hoang Cao
*Conceptual Graphs and Fuzzy Logic,* 2010
ISBN 978-3-642-14086-0

Vol. 307. Anupam Shukla, Ritu Tiwari, and Rahul Kala
*Towards Hybrid and Adaptive Computing,* 2010
ISBN 978-3-642-14343-4

Vol. 308. Roger Nkambou, Jacqueline Bourdeau, and Riichiro Mizoguchi (Eds.)
*Advances in Intelligent Tutoring Systems,* 2010
ISBN 978-3-642-14362-5

Vol. 309. Isabelle Bichindaritz, Lakhmi C. Jain, Sachin Vaidya, and Ashlesha Jain (Eds.)
*Computational Intelligence in Healthcare 4,* 2010
ISBN 978-3-642-14463-9

Vol. 310. Dipti Srinivasan and Lakhmi C. Jain (Eds.)
*Innovations in Multi-Agent Systems and Applications – 1,* 2010
ISBN 978-3-642-14434-9

Vol. 311. Juan D. Velásquez and Lakhmi C. Jain (Eds.)
*Advanced Techniques in Web Intelligence,* 2010
ISBN 978-3-642-14460-8

Vol. 312. Patricia Melin, Janusz Kacprzyk, and Witold Pedrycz (Eds.)
*Soft Computing for Recognition based on Biometrics,* 2010
ISBN 978-3-642-15110-1

Vol. 313. Imre J. Rudas, János Fodor, and Janusz Kacprzyk (Eds.)
*Computational Intelligence in Engineering,* 2010
ISBN 978-3-642-15219-1

Vol. 314. Lorenzo Magnani, Walter Carnielli, and Claudio Pizzi (Eds.)
*Model-Based Reasoning in Science and Technology,* 2010
ISBN 978-3-642-15222-1

Vol. 315. Mohammad Essaaidi, Michele Malgeri, and Costin Badica (Eds.)
*Intelligent Distributed Computing IV,* 2010
ISBN 978-3-642-15210-8

Vol. 316. Philipp Wolfrum
*Information Routing, Correspondence Finding, and Object Recognition in the Brain,* 2010
ISBN 978-3-642-15253-5

Vol. 317. xxx

Vol. 318. Oscar Castillo, Janusz Kacprzyk, and Witold Pedrycz (Eds.)
*Soft Computing for Intelligent Control and Mobile Robotics,* 2010
ISBN 978-3-642-15533-8

Vol. 319. Takayuki Ito, Minjie Zhang, Valentin Robu, Shaheen Fatima, Tokuro Matsuo, and Hirofumi Yamaki (Eds.)
*Innovations in Agent-Based Complex Automated Negotiations,* 2010
ISBN 978-3-642-15611-3

Takayuki Ito, Minjie Zhang, Valentin Robu,
Shaheen Fatima, Tokuro Matsuo, and
Hirofumi Yamaki (Eds.)

# Innovations in Agent-Based Complex Automated Negotiations

Springer

Takayuki Ito
School of Techno-Business Administration
Nagoya Institute of Technology
Gokiso, Showa-ku
Nagoya 466-8555
Japan
E-mail: ito.takayuki@nitech.ac.jp

Minjie Zhang
School of Computer Science and
Software Engineering
The University of Wollongong
NSW 2522, Australia
E-mail: minjie@uow.edu.au

Valentin Robu
School of Electronics and
Computer Science
University of Southampton
SO17 1BJ Southampton, UK
E-mail: Valentin.Robu@cwi.nl

Dr. Shaheen Fatima
Department of Computer Science
Loughborough University
Loughborough LE11 3TU, UK
E-mail: S.S.Fatima@lboro.ac.uk

Tokuro Matsuo
Graduate School of
Science and Engineering
Yamagata University
4-3-16, Jonan
Yonezawa, 992-8510, Japan
E-mail: matsuo@yz.yamagatau.ac.jp

Hirofumi Yamaki
Information Technology Center
Nagoya University
Furo-cho, Chikusa-ku, Nagoya, Japan

ISBN 978-3-642-15611-3 e-ISBN 978-3-642-15612-0

DOI 10.1007/978-3-642-15612-0

Studies in Computational Intelligence ISSN 1860-949X

Library of Congress Control Number: 2010937108



*Typeset & Cover Design:* Scientific Publishing Services Pvt. Ltd., Chennai, India.

Printed on acid-free paper

9 8 7 6 5 4 3 2 1

springer.com

# Preface

Complex Automated Negotiations have been widely studied and are becoming an important, emerging area in the field of Autonomous Agents and Multi-Agent Systems. In general, automated negotiations can be complex, since there are a lot of factors that characterize such negotiations. These factors include the number of issues, dependency between issues, representation of utility, negotiation protocol, negotiation form (bilateral or multi-party), time constraints, etc. Software agents can support automation or simulation of such complex negotiations on the behalf of their owners, and can provide them with adequate bargaining strategies. In many multi-issue bargaining settings, negotiation becomes more than a zero-sum game, so bargaining agents have an incentive to cooperate in order to achieve efficient win-win agreements. Also, in a complex negotiation, there could be multiple issues that are interdependent. Thus, agent's utility will become more complex than simple utility functions. Further, negotiation forms and protocols could be different between bilateral situations and multi-party situations. To realize such a complex automated negotiation, we have to incorporate advanced Artificial Intelligence technologies includes search, CSP, graphical utility models, Bayes nets, auctions, utility graphs, predicting and learning methods. Applications could include e-commerce tools, decision-making support tools, negotiation support tools, collaboration tools, etc.

These issues are explored by researchers from different communities in Autonomous Agents and Multi-Agent systems. They are, for instance, being studied in agent negotiation, multi-issue negotiations, auctions, mechanism design, electronic commerce, voting, secure protocols, matchmaking & brokering, argumentation, and co-operation mechanisms. This book is also edited from some aspects of negotiation researches including theoretical mechanism design of trading based on auctions, allocation mechanisms based on negotiation among multiple agents, case-study and analysis of automated negotiations, data engineering issues in negotiations, and so on. Each paper was carefully reviewed by at least 2 experts in the field.

Finally, we would like to express our sincere thanks to all authors for their hard work. This book would not have been possible without the valuable support and contributions of the cooperators.

Yonezawa, Japan, March 1, 2010

Takayuki Ito
Minjie Zhang
Valentin Robu
Shaheen Fatima
Tokuro Matsuo
Hirofumi Yamaki

# Contents

# Simulation of Sequential Auction Markets Using Priced Options to Reduce Bidder Exposure

Lonneke Mous[1], Valentin Robu[2], and Han La Poutré[3]

[1] CWI, Dutch National Center for Mathematics and Computer Science, Kruislaan 413, NL-1098 SJ Amsterdam, The Netherlands
Erasmus University Rotterdam, Econometrics Institute, P.O. Box 1738, 3000DR Rotterdam, NL

[2] University of Southampton, School of Electronics and Computer Science, Southampton, United Kingdom
vr2@ecs.soton.ac.uk

[3] CWI, Dutch National Center for Mathematics and Computer Science, Kruislaan 413, NL-1098 SJ Amsterdam, The Netherlands
hlp@cwi.nl

**Summary.** This paper studies the benefits of using priced options for solving the exposure problem that bidders with valuation synergies face in sequential auctions. We consider a model in which complementary-valued items are auctioned sequentially by different sellers, who have the choice of either selling their good directly or through a priced option, after fixing its exercise price. We analyze this model from a decision-theoretic perspective and we show, for a setting where the competition is formed by local bidders, that using options can increase the expected profit for both buyers and sellers.

We then perform a comprehensive experimental analysis of our mechanism for different market settings, both with a single synergy bidder, as well as with multiple synergy bidders are active simultaneously. By comparison to our previous work [18, 17], this paper does not focus on analytical results and detailed proofs for the theorems (which are comprehensively reported in Mous et. al. '08 [18]), but it does give more detailed experimental results than reported in previous wok.

## 1 Introduction

Electronic markets represent key coordination mechanisms in multi-agent systems. They allow parties to efficiently allocate resources, tasks and capabilities in large distributed systems, composed of self-interested agents. The rapid rise in online electronic commerce [16, 14], decentralised logistics [31, 21], or even keyword markets in online search [11, 24] have made the development of agent technologies capable of automating such processes increasingly important.

Two main directions of research have emerged in automating such complex agent interactions. One direction of research focuses on designing agents that are able to represent their owners in complex bargaining situations [7, 8, 9, 10], such as negotiating complex contracts in cases where the bargaining agents have non-linear

T. Ito et al. (Eds.): Innovations in Agent-Based Complex Automated Negotiations, SCI 319, pp. 1–25.
springerlink.com 

utility functions over the different issues to be agreed upon [6, 15, 25, 22, 23]. The other main important direction is designing agents that are able to bid in a set of combinatorial, concurrent or sequentially-occurring auctions.

As shown in [1, 3, 20] (among others), a crucial challenge in designing bidding agents that are able to participate in a sequence of auctions is the *exposure problem*. The exposure problem appears whenever a bidder with complementary valuations (i.e. synergies) tries to acquire a bundle of goods sold through sequential auctions. Informally, the problem occurs whenever an agent may buy a single good at a price higher than what it is worth to her, in the hope of obtaining extra value through synergy with another good, which is sold in a later auction. However, if she then fails to buy this other good at a profitable price, she is exposed to the risk of a potential loss. In the analysis presented in this paper, we call such a global bidder a *synergy bidder*.

The exposure problem is well known in auction theory and multi-agent systems research. The usual way to tackle this problem in the mechanism design literature is to replace sequential allocation with a one-shot mechanism, such as a combinatorial auction [27]. However, this approach has the disadvantage of typically requiring a central point of authority, which handles all the bids and payments. Moreover, many allocation problems occurring in practice are inherently decentralized and sequential. Possible examples range from items sold on Ebay by different sellers, loads appearing over time in distributed transportation logistics, dynamic resource allocation in hospitals, etc.

Note that this is a very complex problem, and this paper provides a first decision-theoretic analysis of how priced options can be used to address this problem. However, we do stress options are not a "silver bullet" that completely removes the exposure problem, rather, they are a mechanism that, under some assumptions, removes part of the risk exposure and is preferable to both sides (buyers and sellers), by comparison to a direct sale. In fact, auctions for direct sale of the good (as will become apparent in Section 1.3) becomes, in our option model, a particular sub-case.

## 1.1 Options: Basic Definition

An option can be seen as a contract between the buyer and the seller of a good, subject to the following rules:

- The writer or seller of the option has the *obligation* to sell the good for the *exercise price*, but not the right.
- The holder or buyer of the option has the *right* to buy the good for the *exercise price*, but not the obligation.

Since the buyer gains the right to choose in the future whether or not she wants to buy the good, an option comes with an *option price*, which she has to pay regardless of whether she chooses to exercise the option or not.

Options can thus help a synergy bidder reduce the exposure problem she faces. She still has to pay the option price, but if she fails to complete her desired bundle, then she does not have to pay the exercise price as well and thus she limits her loss. So part of the uncertainty of not winning subsequent auctions is transferred to the seller, who may now miss out on the exercise price if the buyer fails to acquire the desired bundle. At the same time, the seller can also benefit indirectly, from the additional participation in the market by additional synergy bidders, who would have otherwise stayed out, because of the exposure to a potential loss.

## 1.2 Related Work

In existing multi-agent literature, to our knowledge, there has been only limited work to study the use of options.

The first work to introduce an explicit option-based mechanism for sequential-auction allocation of goods to the MAS community is Juda & Parkes [12, 13, 14]. They create a market design in which global bidders are awarded free (i.e. zero-priced) options, in order to cover their exposure problem and, for this setting, they propose truth-telling as a dominant strategy. In their case, the exposure problem is entirely solved for the synergy bidders, because they do not even have a possible loss consisting of the option price. However, this approach also introduces some limitations. First, there may be cases when the market entry effects are not sufficient to motivate the sellers of items to use options. Because the options are assumed to be offered freely (zero-priced), there may be cases in which sellers do not have a sufficient incentive to offer free options, because of the risk of remaining with their items unsold. The sellers could, however, demand a premium (in the form of the option price) to cover their risk. Thus, in such cases, only positively-priced options can provide sufficient incentive for for both sides to use the mechanism. Also, the mechanism described in [13, 14] assumes synergy bidders bid their entire valuation (monetary utility) for their desired bundle on each good of that bundle. This design works with a single synergy bidder - but might not work when several such buyers are active in the market simultaneously.

Priced options have a long history of research in finance (see [5] for an overview). However, the underlying assumption for all financial option pricing models is their dependence on an underlying asset, which has a current, public value that moves independently of the actions of individual agents (e.g. this motion is assumed to be Brownian for Black-Scholes models). This type of assumption does not hold for the online, sequential auctions setting we consider. In our case, each individual synergy bidder has its own private value for the goods/bundles on offer, and bids accordingly.

Another relevant work that studies the use of options in online auctions is that of Gopal et al [2]. Gopal et al. discuss the benefits of using options to increase the expected revenue of a seller of multiple copies of the same good. They do not consider the use of options to solve the exposure problem of buyers with complementary valuations over a bundle of goods (i.e. the synergy bidders in our model). Furthermore, in [2], it is the seller that fixes in advance both the option price and the exercise price, which requires restrictive assumptions on the behaviour of the bidders.

Finally, there is a connection between options and leveled commitment mechanisms [28, 29, 30]. In leveled commitment, typically both parties have the possibility to decommit (i.e. unilaterally break a contract), against paying a pre-agreed decommitment penalty. This differs from option contracts, where the right to exercise the option (i.e. the option price) is paid by one party in advance and is established through an auction.

### 1.3 Outline and Contribution of Our Approach

The goal of this paper is to study the use of priced options to solve the exposure problem and to identify the settings in which using priced options benefits both the synergy bidder and the seller.

An option consists out of two prices, so an adjustment needs to be made to the standard auction with bids of a single price. The essence of options, in our model, is that buyers obtain the right to buy the good for a certain exercise price in the future. The value of such an option may be different for different market participants at different times. Throughout this study, in order to make the analysis tractable, we have a fixed exercise price and a flexible option price. The seller determines the exercise price of an option for the good she has for sale and then sells this option through a first price auction. Buyers bid for the right to buy this option, i.e. they bid on the option price.

Note that, in this model, direct auctioning of the items appears as a particular subcase of the proposed mechanism, assuming free disposal on the part of the buyers. If the seller fixes the future exercise price for the option at zero, then a buyer basically bids for the right to get the item for free. Since such an option is always exercised (assuming free disposal), this is basically equivalent to auctioning the item itself.

Based on the above description, we provide both an analytical and an experimental investigation of the setting. Our analysis of the problem can be characterized as decision-theoretic, meaning both buyer and seller reason with respect to expected future price. In summary, our contribution to the literature can be characterized as being twofold:

First, we consider a setting in which $n$ complementary-valued goods (or options for them) are auctioned sequentially, assuming there is only one synergy bidder or global bidder (the rest of the competition is formed by local bidders desiring only one good). For this setting, we show analytically (under some assumptions), that using priced options can increase the expected profit for both the synergy bidder and the seller, compared to the case when the goods are auctioned directly. Furthermore, we derive the equations that provide minimum and maximum bounds between which the bids of the synergy bidder are expected to fall, in order for both sides to have an incentive to use options.

In the second part of the paper, we consider market settings in which multiple synergy bidders (i.e. global bidders) are active simultaneously, and study it through experimental simulations. In such settings, we show that, while some synergy bidders lose because of the extra competition, other synergy bidders may actually benefit, because sellers are forced to fix exercise prices for options at levels which encourages participation of all buyers.

The structure for the rest of this paper is as follows. Section 2 lays the foundation, and derives the expected profits of synergy bidders and sellers for both the direct sale, respectively for a sale with options. Section 3 provides the outline behind the analytical results and proofs, for a market of sequential auctions with one synergy bidder. It is important to note that this paper does *not* provide the detailed proofs (interested readers can find this in [18]), but we do provide enough ingredients, such that the reader can understand the basic ideas behind our model and analysis. Then, Sections 4 and 5 give the results from our experimental investigations for multiple market settings, while Section 6 concludes with a discussion.

## 2 Expected Profit for a Sequence of $n$ Auctions and One Synergy Bidder

In this section, we provide the basis behind the analytical results of our work, by deriving the expected profit functions for both the synergy bidder and the seller in our model. Afterwards, in Section 3 we will give a sketch of the analytical proof that options can be profitable for both the synergy bidder and the auctioneer in this setting (this is only a sketch because, as explained in the introduction, this paper does not emphasize the analytical part which is fully discussed in other work by the same authors [18]). Important to note that, throughout this study, it is assumed that both sellers and buyers are risk neutral and that they want to maximize their expected utility or - in this case - their expected profit.

### 2.1 Profit with $n$ Unique Goods without Options

This section describes the expected profit of the synergy bidder and the sellers as a function of the synergy bidder's bids for a market with $n$ unique, complementary goods, which are sold without options.

Let $G$ be the set of $n$ goods for sale in a temporal sequence of auctions and $v_{syn}(G_{sub})$ be the valuation the synergy bidder has for $G_{sub} \subseteq G$. Then assume that $v_{syn}(G) > 0$ and $\forall G_{sub} \subsetneq G,\ v_{syn}(G_{sub}) = 0$. In other words, the synergy bidder only desires a bundle of all the goods considered in the model.

The goods $G_1..G_n \in G$ are sold individually through sequential, first-price, sealed-bid auctions. Here we choose the auctions to be first price, as they are more tractable to study using game-theoretic analysis. Furthermore, in a sequential setting with valuation complementarities of the agents, second-price auctions do not have the nice dominant strategies properties, described by Vickrey. Furthermore, in many settings where such a model could be used in practice, such as request-for-quotes (RFQ) auctions in logistics or supply chains, first-price auctioning is often used.

The time these auctions take place in is $t = 1 \dots n$, such that at time $t$ good $G_t \in G$ is auctioned. The above assumptions mean that if the synergy bidder has failed to obtain $G_t$, then she cannot achieve a bundle, for which she has a positive valuation. So if $G_{t+1}$ is auctioned with a positive reserve price, then obtaining $G_{t+1}$ will only cost the synergy bidder money. Therefore, if the synergy bidder fails to obtain $G_t$, then it is rational for her to not place bids in subsequent auctions.

The bids of the synergy bidder are $\mathbf{B} = (b_1, \dots, b_n)$, where $b_t$ is the bid the synergy bidder will place for good $G_t$, conditional on having won the previous auctions.

Because of the first-price auction format, $b_t$ is also the price the synergy bidder has to pay if she has won the auction.

Throughout this analysis, we assume the competition the synergy bidder faces for each good $G_t$ (sold at time $t$) is formed by local bidders that only require the good $G_t$. We further assume that these local bidders are myopic, i.e. the bids placed by the synergy bidder have no effect on their bidding behaviour. Therefore, from the perspective of the synergy bidder, the competition can be modeled as a distribution over the expected closing prices at each time point $t$, more precisely as a distribution over a value $bm_t$, which is the maximal bid placed by the competition not counting $b_t$.

Denote by $F_t(b_t)$ the probability that the synergy bidder wins good $G_t$ with bid $b_t$ - where $F_t(b_t)$ depends on whether $b_t$ can outbid the maximal bid $bm_t$ placed by the competition, excluding $b_t$. For each good $G_t$, there exists a strictly positive reserve price of $b_{t,res}$, which is the seller's own valuation for that good. Then $bm_t$ is the highest bid of the local bidders (who only want $G_t$), if that bid is higher than $b_{t,res}$. Otherwise $bm_t$ equals $b_{t,res}$. To deal with ties, we assume the synergy bidder only wins $G_t$ if $b_t > bm_t$ and not if the bids are equal. Then $F_t(b_t)$ can be defined as follows:

$$F_t(b_t) = Prob(b_t > bm_t) \tag{1}$$

The synergy bidder only has a strictly positive valuation for the bundle of goods $G$, which includes all the goods $G_t$, sold at times $t = 1..n$.

$$E(\pi^{dir}_{syn}) = \left[v_{syn}(G)\prod_{i=1}^{n} F_i(b_i)\right] + \left[\sum_{j=1}^{n}(-b_j)\prod_{k=1}^{j} F_k(b_k)\right] \tag{2}$$

The synergy bidder wants to maximize her expected profit. So her optimal bids $\mathbf{B}^* = (b_1^*, \ldots, b_n^*)$ maximize equation 2:

$$\mathbf{B}^* = argmax_{\mathbf{B}^*}\ E(\pi^{dir}_{syn}) \tag{3}$$

Next the profit of the sellers are examined. It is assumed that all sellers have their own valuation for the good that they sell and that they set their reserve price of $b_{t,res}$ equal to this private valuation. So when the good is sold for $b_t$, the seller of $G_t$ has a profit $\pi^{dir}_t$ of $b_t - b_{t,res}$. As previously shown, the synergy bidder only participates when she has won the previous auctions; otherwise $bm_t$ is the maximal placed bid. The expected profit of the seller of the good $G_t$ sold at time $t$ is:

$$E(\pi^{dir}_t) = (E(bm_t) - b_{t,res})(1 - \prod_{i=1}^{t-1} F_i(b_i)) + \Big(F_t(b_t)(b_t - b_{t,res}) + (1 - F_t(b_t))(E(bm_t|bm_t \geq b_t) - b_{t,res})\Big)\prod_{i=1}^{t-1} F_i(b_i) \tag{4}$$

Intuitively explained, the equation defines the expected utility over 3 disjoint cases: one in which the optimal bids $b_i$ of the synergy bidder were sufficient to win all auctions up to time $t$, in which case the expected profit of the seller is the highest expected bid of the local bidders $E(bm_t)$, minus its own reservation value $b_{t,res}$; the second case in which the synergy bidder wins all previous auctions, including the current one (i.e. the one at time $t$), in which case the expected profit is this bid minus reservation $b_t - b_{t_res}$, and the third in which the synergy bidder won all previous auctions but fails to win the current one, in which case still the highest bid by the local bidders is taken.

## 2.2 Profit with $n$ Unique Goods with Options

Section 2.1 derived the expected profit functions for the synergy bidder and the sellers in a market without options. The next step is to do the same for a market with options. This section has the same setting as the general model with $n$ goods being sold, only now an option on $G_t$ is auctioned at time $t$. Therefore, all the sellers in the market will sell options for their goods, instead of directly the goods themselves. After the $n$ auctions have taken place, the buyers need to determine whether or not they will exercise their option. It is assumed that an option is only exercised if a buyer has obtained her entire, desired bundle. The local bidders are only interested in $G_t$, so they will always exercise an option on $G_t$ should they have one. The synergy bidder is only interested in a bundle of all goods, so she will only exercise an option (and pay the corresponding exercise price) if she has options on all the goods required.

The option exists out of a fixed exercise price $K_t$ and the synergy bidder's bids on the option price are $\mathbf{OP} = (op_1, \ldots, op_n)$. The maximal bid without the synergy bidder was $bm_t$, but now $opm_t$ is the maximal placed option price.

Since the competition only wants one good, they do not benefit from having an option and they will always exercise any option they acquire. Therefore the competition's best policy is to keep bidding the same total price, which is the bid without options minus the exercise price. Thus the distribution of the competition is only shifted horizontally to the left, by the reduction of the exercise price: $opm_t = bm_t - K_t$. Thus, if the synergy bidder bids the same total price (option + exercise), then she has the same probability of winning the auction in both models. Let $F_t^o(op_t)$ be the probability that $op_t$ wins the auction for the option on $G_t$. So if $op_t + K_t = b_t$, then $F_t^o(op_t) = F_t^o(b_t - K_t) = F_t(b_t)$.

The synergy bidder's expected profit with options then is:

$$E(\pi_{syn}^{op}) = \left[ \left(v_{syn}(G) - \left[\sum_{h=1}^{n} K_h\right]\right] \prod_{i=1}^{n} F_i^o(op_i) + \left[\sum_{j=1}^{n}(-op_j) \prod_{k=1}^{j} F_k^o(op_k)\right] \tag{5}$$

So her optimal bids $\mathbf{OP}^* = (op_1^*, \ldots, op_n^*)$ maximize the profit equation 5:

$$\mathbf{OP}^* = argmax_{\mathbf{OP}^*} \; E(\pi_{syn}^{op})) \tag{6}$$

The main difference for the seller of $G_t$, is that if the synergy bidder wins, then she only earns $K_t - b_{t,res}$ when the option is exercised. She then gains the exercise price, but loses the value the good has to her, which is the reserve price. And the probability of exercise is the probability that the synergy bidder wins all the other auctions. Therefore, the total expected profit of the seller at time $t$ is:

$$E(\pi_t^{op}) = (E(opm_t) + K_t - b_{t,res})(1 - \prod_{i=1}^{t-1} F_i^o(op_i))$$
$$+ \Big(F_t^o(op_t)(op_t + \Big[(K_t - b_{t,res}) \prod_{h=t+1}^{n} F_h^o(op_h)\Big])$$
$$+ (1 - F_t^o(op_t))(E(opm_t|opm_t \geq op_t) + K_t - b_{t,res})\Big) \prod_{i=1}^{t-1} F_i^o(op_i) \quad (7)$$

Briefly explained, this equation has the same 3-case structure as Eq. 4 above. In two cases: when the synergy bidder loses an auction for one the earlier items in the sequence (before the items sold at time $t$), or when she wins all the earlier auctions, but not the auction at time $t$, the expected payoffs are equivalents to the direct auctioning case, although this time expressed slightly differently, based on both the exercise and option price. However in one case, when the synergy bidder acquires all the previous items and the current one (middle line in Eq. 7), the payoff is composed of two amounts. The option price $op_t$ will be gained for sure, in this case. However, the difference between the exercise and reserve price $K_t - b_{t,res}$ (which signifies the item actually changes hands) is acquired only if the synergy bidder also wins all the subsequent auctions at times $h = t + 1..n$. This is an important difference, and it would seem from these equations that the seller has no interest to use options, since in one important case, part of the amount she is about to receive depends on the outcome of future auctions. The key, however, rests in the observation that the synergy bidder should be willing to bid more in total (i.e. $K_t + op_t$) than in the direct auctioning case. This will be analyzed in the next Section.

## 3 When Options Can Benefit Both Synergy Bidder and Seller

Section 2 resulted in the a-priori, expected profit for the synergy bidder and the sellers as a function of the synergy bidder's bids for a market with and without options. This section uses these functions to determine the difference in profit between the two markets, which is $\pi_{\delta t}$ and $\pi_{\delta syn}$ for the seller of good $G_t$ and the synergy bidder respectively, where:

$$\pi_{\delta t} = \pi_t^{op} - \pi_t^{dir},$$
$$\pi_{\delta syn} = \pi_{syn}^{op} - \pi_{syn}^{dir}$$

So if $\pi_{\delta t}$ and $\pi_{\delta syn}$ are positive, then both agents are better off with options.

### 3.1 When Agents Are Better Off with Options

Let $\mathbf{B}^*$ denote the synergy bidder's optimal bidding policy in a market where goods are sold directly (without options). We assume for the rest of Sect. 3 that for $1 \leq t \leq n$, $F_t(b_t^*) > 0$ and $F_t(b_t^*) < 1$. So she may complete her bundle, but may also end up paying for a worthless subset of goods. Thus she faces an exposure problem. For the market with options, we define a benchmark strategy $\mathbf{OP}'$ for the synergy bidder, so that the two markets can easily be compared. The benchmark of the synergy bidder's bids with options $\mathbf{OP}' = (op'_1, \ldots, op'_n)$ is that for $1 \leq t \leq n$:

$$op'_t = b_t^* - K_t$$

In other words, the benchmark strategy implies that the synergy bidder will bid the same total amount for the good, as if she used her optimal bidding policy in a direct sale market. Clearly this does not have to be her profit-maximizing bid in a market where priced options are used. In fact, it is almost always the case that the synergy bidder will bid a different value in a market in with priced options. This deviance from the benchmark is denoted by $\lambda_t$:

Let $\lambda_t$ denote the deviation in the bid of the synergy bidder on the item $G_t$ sold at time $t$, in a model with options, with respect to her profit-maximizing bid $b_t^*$ in a model without options. So her bid on an option for $G_t$ will be $op'_t + \lambda_t$.

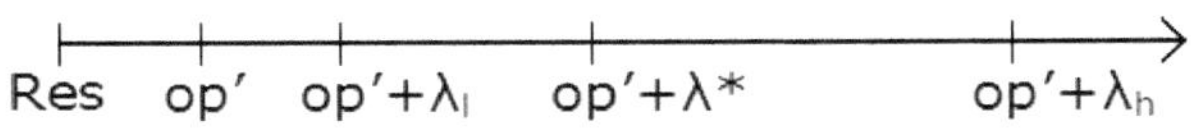


**Fig. 1.** A possible situation in which options are desirable.

These definitions enable us to rigorously define the bounds within which the use of options (with a given exercise price) are desirable for both the synergy bidder and the seller, for each good in the auction sequence (except the last one, for which there is no uncertainty, so the use of options is indifferent). Fig. 1 gives the visual description of a generic setting in which options are beneficial for both sides. It shows the possible bids a synergy bidder can place for an option. First, valid bids have to be bigger than the reserve price $Res$, for each good in the sequence. The point $op'$ is where the synergy bidder keeps bidding the same total price as in a market without options, as in the definition of $op'$ above. The deviations, in an option model, from the benchmark bid $op'$ is measured by three levels, all denoted with $\lambda$: $\lambda_l$ is the minimal risk premium the seller requires to benefit from using options, $\lambda_h$ is the maximal extra amount the synergy bidder is willing to pay for an option and $op^* = op' + \lambda^*$ is the synergy bidder's profit-maximizing bid in an option market. So, if it is rational for the synergy bidder to bid an additional quantity between $\lambda_l$ and $\lambda_h$ (as shown in Fig. 1), then both she and the seller are better off with options.

As mentioned in the abstract and introduction, we will not analytically derive the equations that give these levels of $\lambda$-s above (interested readers can consult [18] for more details). In this paper, we focus our attention on the different test we performed for different sequential auction markets, with either a single or multiple synergy bidders. For all case, we experimentally compare expected profits for markets with direct sale vs. option markets.

## 4 Simulation of a Market with a Single Synergy Bidder

This section presents an experimental examination of a market with one synergy bidder. It introduces the market entry effects in the synergy bidder's behaviour, as well as the threshold effects that may determine which exercise prices the seller chooses for her options. This experimental analysis is performed here for a market with one synergy bidder and several local bidders, while Sect. 5 considers a market with multiple synergy bidders.

The experimental setting is as follows: we consider a simulation where two goods A and B are auctioned $n_A$ and $n_B$ times respectively. The synergy bidder desires one copy of both goods and has zero valuation for the individual goods. That is, each synergy (or global) bidder requires exactly one bundle of $\{A, B\}$[1] In the setting considered in this Section, local bidders only want one good and participate in one auction, thus their bids can be modeled as a distribution.

Furthermore, in order to simplify the simulation we assume there is a single seller who auctions all the goods. This is actually equivalent to studying whether *on average* sellers have an incentive to use options. To explain, on any single sequence of auctions taken in isolation, the sellers of different items may have highly diverging incentives to use options, based on their position in the auction queue. However, in a very large setting, where buyers enter the market randomly, it is difficult for any individual seller to strategise about her particular place in the sequence (and, furthermore, in most markets she may simply have no information to do this). Our goal is to study under which conditions, on average, sellers benefit from using options if there are synergy bidders in the market. Also, to somewhat reduce the number of test parameters, we further assume that the exercise price is the same for all goods of the same type. So the seller needs to determine which exercise price for A and which for B maximize her expected profit.

Note that, typically a seller has a resale value of for the goods that remain unsold, which is typically lower that the value at the start of the auction sequence. The reason for this may be that there is some time discounting associated with waiting for a sequence of auctions to resell her items, or even a listing cost, which is paid per auction (such as in the Ebay case). In this paper, we do not explicitly simulate resale, but we use a reservation value, which represents the expected resale value the seller expects to get, if she is forced to resell her items.

[1] An intuitive way to think about this setting is as a sequential sale of individual shoes of exactly the same type, where $A$ is the left shoe, and $B$ is the right shoe, and each synergy bidder requires exactly one pair.

To summarize, simulations were run in Matlab and had the following parameters:

| Name | Explanation |
|---|---|
| $n$ | The number of auctions. |
| $mean$ | The mean of price distribution. |
| $std$ | The standard deviation of price distribution. |
| $res$ | Reserve prices. |
| $v_{syn}$ | Valuation synergy bidder for A and B combined. |
| $k$ | Number of simulations for each auction run |

A basic simulation run is as follows. First, all possible auction sequences are determined for the given number of auctions for A and B. The simulation is then run for all these sequences, both for a direct sale setting and for a setting where the items are sold through options with given exercise prices.

For each auction, in each simulation run, there is a set of local bidders, assumed myopic. The bids of these local bidders are therefore, assumed to follow a normal price distribution, with the parameters $n, mean, std$ and $res$ consisting out of two values: one for good A and one for good B. For each simulation run, the synergy bidders(s) are asked to determine their profit-maximizing bid for that setting, as described in the next section. The optimization required for determining their optimal bid is done using the Matlab function "fminsearch" from the Optimization Toolbox.

Since there may be considerable variance in the bids of the local bidders (which are myopic) each possible auction sequence is run $k$ times (typically, we had $k > 10000$). The average profit of the seller and the synergy bidder which are reported here, for both the case of with and without options, are averages over all these $k$ simulations and also over all possible auction orders of items A and B in the sequence.

## 4.1 Deriving the Optimal Bid Strategy of the Synergy Bidder

This section describes how the synergy bidder determines her bids in the simulation. In order to neutralize the effect that the exact order items are auctioned in plays on the bidding strategy, we add the assumption that the synergy bidder knows the number of remaining auctions, but not the order they will be held in. This remaining number of auctions of each type is common knowledge (i.e. the synergy bidders can always observe how many auctions of each type are left before they have to leave the market, and so does the seller).

The model described here is for a situation without options. But in order to apply it to a situation with options, one merely has to replace the variables: $b_t = op_t - K_t$ and $v_{syn}(A, B) := v_{syn}(A, B) - K_A - K_B$. As in the analytical section, we assume a bidder only wants a complete bundle of $\{A, B\}$. Therefore, $v_{syn}(A) = v_{syn}(B) = 0$.

Determining the synergy bidder's profit-maximizing bid $b_t^*$ at state $t$ basically involves solving the Markov Decision Process (MDP), where we select the optimal bid $b_t^*$ at time $t$, subject to the optimal bid $b_{t+1}^*$ being selected for the future time point $t + 1$ (which in this case, is an auction). We can, however, use the valuation function of the bidding agent to significantly reduce the state space of the MDP, as shown below. However, first we introduce some notation.

Let $b^*$ be the immediate best response to the state, which depends on four variables: $z_A, z_B, X$ and $I_t$. The variables $z_A$ and $z_B$ are the number of remaining auctions for $A$ and $B$ respectively (including the current auction), so $z_A \leq n_A$, $z_B \leq n_B$. The type of good, which is currently sold, is denoted by $I_t$. The set of goods the synergy bidder owns (i.e. the endowment) is described by $X$, which can either be $\emptyset, \{A\}$ or $\{B\}$. If $X$ is $\{A, B\}$ then the synergy bidder is done. Let $Q(z_A, z_B, X, I_t, b_t)$ be the expected profit of the synergy bidder when bidding $b_t$. Note that, in these definitions, $b^*_{t+1}$ and $V_{t+1}()$ denote the best available bid, respectively best expected value for the next state (as computed by recursion), while $I_{t+1}$ is the type of the next item in the auction sequence. Therefore, using MDP notation, the profit-maximizing bid $b^*_t$ is determined as follows:

$$b^*_t = argmax_{b_t}\ Q(z_A, z_B, X, I_t, b_t) \tag{8}$$

Where the expected profit is determined via:

$$\begin{aligned} &Q(z_A, z_B, X, I_t = A, b^*_{t+1}) = F_A(b_t)(-b_t \\ &+ V_{t+1}(z_A - 1, z_B, X \cup A, b^*_{t+1}))+ \\ &(1 - F_A(b_t))V_{t+1}(z_A - 1, z_B, X, b^*_{t+1}) \end{aligned} \tag{9}$$

$$\begin{aligned} &Q(z_A, z_B, X, I_t = B, b_t) = F_B(b_t)(-b_t \\ &+ V_{t+1}(z_A, z_B - 1, X \cup B, b^*_{t+1}))+ \\ &(1 - F_B(b_t))V_{t+1}(z_A, z_B - 1, X, b^*_{t+1}) \end{aligned} \tag{10}$$

Where $V()$ is the value of a state, which simply means the maximum expected profit of that state:

$$V_t(z_A, z_B, X, b_t) = max_{b_t}\ Q(z_A, z_B, X, I_t, b_t) \tag{11}$$

Looking at the formula for $Q()$, it basically says that for the probability of winning the auction with her bid, the synergy bidder has to pay a price equal to her bid and the good is included in the endowment $X$ of the next state. If she does not win the auction, then the value of the current state is equal to the value of the next state.

As we mentioned before, in computing its optimal bidding strategy used in the experimental Section, we assume the synergy bidder does not know whether the next auction will be for A or B, she only knows the total numbers of auctions for A and B remaining. We acknowledge this is a departure from the formulas in the theoretical analysis, where the exact order of the auctions was taken into account to compute the bidding strategies. There are two reasons to use this assumption here. The first is that it reduces considerable the state space that needs to be modeled when computed the optimization. But the second is that we also find this choice more realistic if this model is to be applied to real-life settings. For example, when bidding on a part-truck order in a logistic scenario, it is more realistic to assume that a carrier can approximate the number of future opportunities to buy a complementary load, but not the exact auction order in which future loads will be offered for auction.

If we assume the synergy bidder only knows the total numbers of auctions for A and B remaining (and not their exact order), then her bidding strategy is based on assuming each future auction has an equal probability to occur. Therefore, the probability of an auction for A occurring next is simply the number of remaining auctions A divided by the total number of remaining auctions. Thus, a weighted average can be used to determine the value of the next auction, while not knowing for which good it will be for.

Apart from this general framework, we can prune the state space with the cases in which we know the synergy bidder's bid is zero:

$$b_t^* = argmax_{b_t}\ Q(0, z_B, X, B, b_t) = 0, \text{ with } A \notin X \tag{12}$$
$$b_t^* = argmax_{b_t}\ Q(z_A, 0, X, A, b_t) = 0, \text{ with } B \notin X \tag{13}$$
$$b^t * = argmax_{b_t}\ Q(z_A, z_B, X, I_t \in X, b_t) = 0 \tag{14}$$

With the first two cases, the synergy bidder can no longer obtain her desired bundle, because she does not own the complementary item and there is no chance left of acquiring it. The last equation is for the case when the synergy bidder already has a copy of the type of good (and, from her valuation function, she only wants exactly one copy of A and B). The corresponding values of these states are:

$$V(0, z_B, X, b_t^*) = 0, \text{ if } A \notin X(z_A = 0, asI_{t+1} = B) \tag{15}$$
$$V(z_A, 0, X, b_t^*) = 0, \text{ if } B \notin X(z_B = 0, asI_{t+1} = A) \tag{16}$$
$$V(z_A, z_B, \{A\}, b_t^*) = V(0, z_B, \{A\}, b_t^*) \tag{17}$$
$$V(z_A, z_B, \{B\}, b_t^*) = V(z_A, 0, \{B\}, b_t^*) \tag{18}$$

The first two equations correspond to the case when the buyer can no longer get the complementary-valued item, therefore the sequence of auctions of the same type has no value to her. In both these cases $b_t^* = 0$. The last two equations are important, since they help the most to reduce the state space. Basically, as already mentioned, we assume that a synergy bidder only wants exactly one bundle of $\{A, B\}$. If she already owns a good of one of the two types, she will no longer be interested in the remaining auctions for that type of good. Therefore, the valuation $V()$ of these states is equivalent to a state when no auctions are remaining for the type of good she already owns (as she would not take part in those anyway). All these techniques help reduce the recursive search.

To conclude, to determine the synergy bidder's bids in any situation, the values of $b_t^*$ and $V()$ need to be calculated for the following states:

$$\begin{aligned}
\forall z_B > 0 \quad & Q(0, z_B, \{A\}, B, b_t) \\
\forall z_A > 0 \quad & Q(z_A, 0, \{B\}, A, b_t) \\
\forall z_A > 0, z_B > 0 \quad & Q(z_A, z_B, \emptyset, A, b_t) \\
\forall z_A > 0, z_B > 0 \quad & Q(z_A, z_B, \emptyset, B, b_t)
\end{aligned}$$

### 4.2 Experimental Results: Market Entry Effect for One Synergy Bidder

First, we study experimentally the incentives to use options for the sellers and buyers, in the case there is just one synergy bidder present in the market. In order to study different dimensions of such markets, we considered several combinations of parameter settings.

The first setting has $n_A = 2$ and $n_B = 2$. As mentioned above, the local bidders are considered myopic and only bid in one local auction. Therefore, their bids can be modeled as a distribution $\sim N(10, 4)$ for both goods. The goods A and B are, in this model, of equal rarity and attract an equal amount of independent competition during bidding. This choice is not random, as having a certain degree of symmetry in the experimental model allows us to reduce the number of parameter settings we need to consider. More specifically, we assume the same exercise prices are set for both goods of type A and B. This is a reasonable assumption, because A and B are of symmetric value and because bidders do not know in advance the exact order goods will be sold in.

Furthermore, for each good, the seller has a reservation value $res = 8$, which gives its estimate resell value in the case the synergy bidder acquires an option for the item, but fails to exercise it. Since, on average, myopic bidders bid have an expected mean of 10 for an item, 20% is a reasonably safe estimate of a resell value.

The value of a bundle of {A,B} for the synergy bidder is an important choice, especially in relation to the mean expectation $\mu$ of the bids placed by single-item bidders. We considered two settings: $v(A, B) = 24$ (thus 20% more, on average, than local competition) - with results shown in Fig. 2, and $v(A, B) = 21$ (which is only 5% more on average than local competition) - with results shown in Fig. 3.

Looking at these two figures, some important effect can be observed. First, we mention that the seller has an immediately higher expected profit with options compared to direct sale. This is because an option is sometimes not exercised and then the seller gets to keep the good (for which she has a positive valuation), while the synergy bidder still pays the option price.

There are two main effects to be observed from Fig. 2 and 3:

- First, the synergy bidder in such a market always prefers *higher* exercise prices (an effect clearly seen in both Figs. 2 and 3). This may be counter-intuitive at first, but is a rational expectation. If the option for an item is sold with a higher exercise price, then the synergy bidder can bid more aggressively on the option price to get the item, since she is "covered" for the loss represented by the exercise price. The myopic bidders extract no advantage from being offered the good as an options vs. a direct sale, because, if they acquire the option, they would always exercise it regardless. Therefore, they will simply lower their bid for the option with the amount represented by the exercise price.
- Second, the expected profit of the seller seems to decrease between intervals if she has to sell the option with a higher exercise price. The main reason for this is that there is some chance that she or she would remain with her item unsold (because the option is not exercised), and thus only extract her reservation value for that item. There is, however, an important difference between the cases shown

**Fig. 2.** Percentage increase in profit for a model using options wrt. direct sale, for the case there is one synergy bidder is present in the market. In the setting, there are two items of type A sold and two items of type B. For all 4 items, the bids of the local bidders follow the distribution $N(10, 4)$, while the valuation of the synergy bidder is $v(A, B) = 24$ (thus 20% more, on average, than the local bidders). What is varied on the horizontal axis is the exercise price with which the items are sold (assuming they are set the same for all items, being of equal rarity). Note that the figure is super-imposed: the left-hand side axis refers exclusively to the seller, while the right-hand side axis refers exclusively to the synergy bidder. From this picture, one can already see the important effect: synergy bidder prefers, on average, higher exercise prices, while seller prefers lower ones.

in Fig. 2 and 3, which is the participation thresholds (that appear as "peaks" in the picture), where the expected profit of the seller seems to "jump" at a new level. These can be explained by the synergy bidder joining the market, as the expected profit becomes non-negative. The threshold nature is determined by the discrete nature of the auction sequence, as is explained below.

Such a participation threshold is illustrated in Fig. 3 is the increase in the seller's expected profit when the exercise price is set above a certain level ($K \geq 2.5$, for the settings in Fig. 3). Such thresholds can be explained as follows. If the synergy bidder currently owns nothing, then she will only bid on a good if the number of remaining auctions and their exercise prices give her a prior expectation of a positive profit. Conversely, if the synergy bidder is not offered a sequence of option sales from which she derives a positive expected profit, she has the incentive to leave the market altogether. There are two main factors that increase a synergy bidder's expected profit in a sequence of auctions (sold as options):

- The number of remaining future auctions of the other good, necessary to complete her bundle.

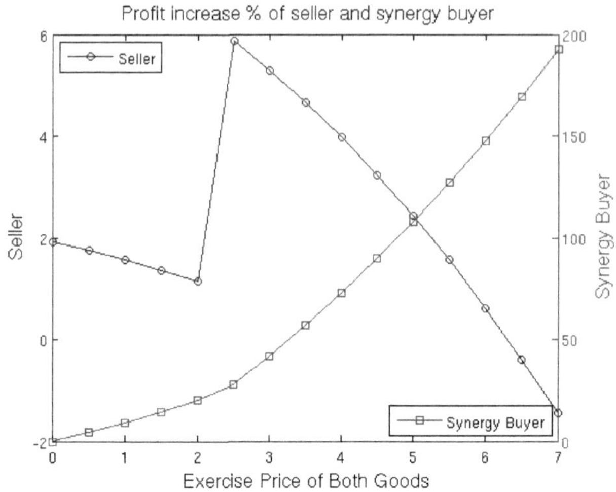


**Fig. 3.** Percentage increase in profit for a model using options wrt. direct sale, for the case there is one synergy bidder is present in the market. The settings are exactly the same as those is in Fig. 2 above: 2 auctions for A and 2 for B, with local, myopic bidders following $N(10,4)$. However, now the valuation of the synergy bidder is $v(A,B) = 21$ (thus only 5% more, on average, than the local bidders). One can see, however, that there is an important difference by comparison to Fig. 2: the threshold effect in the profit increase for the seller when the exercise price $K \geq 2.5$. Intuitively, the reason this effect occurs is the market-entry effect on the part of the synergy bidder, who would otherwise stay out for this lower valuation

- The exercise price of the options (that only needs to be paid at the end). This should be high enough to cover the risk, given her valuation for the bundle.

Note that in some market setting (such as the one in Fig. 2), no participation effects (i.e. thresholds) occur, because the value the synergy bidder assigns to her desired bundle is already high enough, so she would participate in the market anyway (i.e. regardless of whether she gets offered options or not), and at any point in the sequence that there is still a chance of completing her bundle.

However, in the valuation settings in Fig. 3, the synergy bidder will only bid on a good if there are two remaining auctions for the other good. So she places a bid for A if the auctions are $[A, B, B]$, but not if they are $[A, B]$. This is because with a single auction for B, the risk of ending up with a only a worthless A is too great. But in a market with exercise prices of at least 2.5, the risk is reduced and one remaining auction is already enough for the synergy bidder to stay in the market. So a higher exercise price enables the synergy bidder to stay the market, even if she owns nothing and there are only a few auctions left, which increases the seller's expected profit. This increase in participation is beneficial to the seller, who thus has an incentive to fix the exercise prices $K_A = K_B = 2.5$.

### 4.3 Settings with Longer Sequences of Auctions and Effect of Auction Order

In the previous Section, we examined a sequence of auctions of a specific length of $n_A = 1, n_B = 2$. We now look at whether we can observe similar effects in the case when the number of opportunities to buy goods A and B increases. With the exception of auction lengths, the parameters are kept the same as in the previous case. First, we keep the relative rarity of both goods symmetrical, but increase the number of auctions available for each to 4, i.e. $n_A = n_B = 4$. Results are shown in Fig. 4.

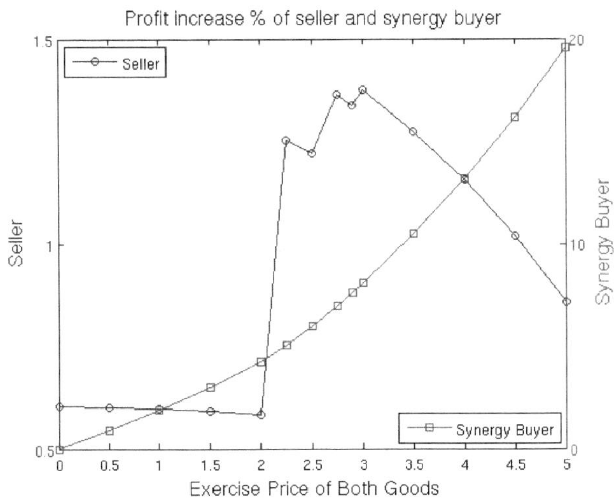


**Fig. 4.** Percentage increase in profit for the case of one synergy bidder, for longer auction sequences. The settings in terms of valuations are exactly the same as those is in Fig. 3 above: the synergy bidder has a value $v(A, B) = 21$, while single-item bidders bid according to $N(10, 4)$. One change is that now there are 4 auctions available for each type, i.e. 4 auctions for an item of type A and 4 for B. Notice that now there are multiple thresholds, since there are multiple points when the market entry effect of the synergy bidders appears. However, on average, the percentage increases in expected profits for the synergy bidders are lower, when compared to the direct auctions case. The reason for this is that, with multiple future buying opportunities, the exposure problems that synergy bidder faces decreases.

Basically, there are two main effects to observe here. First, the benefits to the buyer of having options mechanism decreases (seen from comparing the percentage increases shown in the right-hand vertical axis of Figs. 3 and 4). The reason for this (as discussed in the earlier, risk-based bidding paper) is that, in sequential auctions, the number of available future opportunities plays a big role in how big the exposure problem the synergy bidder faces is. If there is less exposure, then the relative benefits of using options becomes smaller (although it is still quite considerable). The second effect to be observed from Fig. 4 is that there are more participation

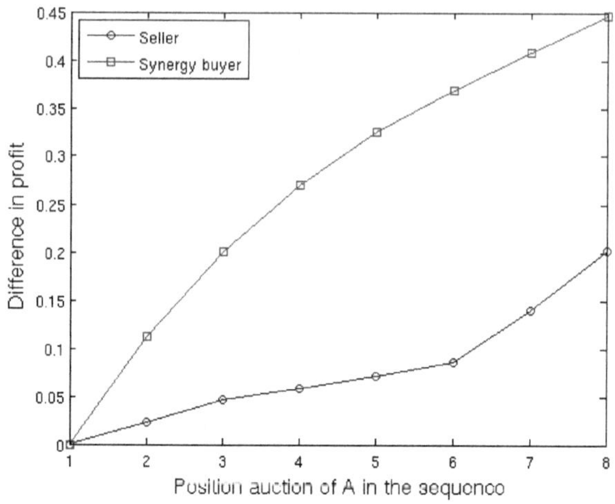


**Fig. 5.** Influence of the position in an auction queue of an item on the seller's expected profit. Settings are the same as in Fig. 2, but with one important difference: the rarity of the goods is no longer symmetric. There is now only 1 auction for a good of type A, but 7 auctions for a good of type B. What is varied along the horizontal axis is the position in the auction queue of the sale of the rarer item (of type A). The graph shows the absolute difference in profit for a seller of an item *of type B* and for the synergy bidder (i.e. the difference in profit between an options and direct auctions model). Note that, if the rare item of type A is sold at the end of the auction sequence, the benefit of selling item B through an option increases, because the exposure risk of not acquiring item of type A increases.

thresholds (denoted by peaks), but they are smaller. The reason is that, for a longer sequence of auctions, there are more possible sequences of remaining auction combinations. The synergy bidder will join in the bidding in some, but not in others, leading to multiple participation thresholds.

The second problem we look in this subsection at is what happens if the relative frequency of the two goods is more asymmetric. We keep the same total number of auctions in the sequence (8), but the relative frequency is highly asymmetric: $n_A = 1, n_B = 7$. In the previous graphs, results were averaged over all possible auction orders - while here, by contrast, we look at auction orders one by one.

For this setting, there are exactly 8 possible auction orders, corresponding to the point where the rarer good (type A) can be inserted in the auction queue. What is varied on the horizontal axis is this position of the type $A$ good. The reason why we look at whether a seller of items of type $B$ would use options is that the exposure of the synergy bidder exists for the other good in the sequence. For the single item of type $A$, the benefits of using options are limited, because the synergy bidder has 7 other auctions in which to acquire the second item anyway, hence she has much less of an exposure problem. Results are plotted in Fig. 5.

Clearly, we can see an important effect of the position of the rarer good in the auction queue, from the perspective of both parties. If the item of type A is sold at the very beginning of the auction sequence, then the synergy bidder has no exposure problem left for the rest of the sequence, hence there is no incentive to use options, for either party. However, it is at the very end of the auction sequence, the synergy bidder will not know whether she would need the item acquired until all auctions end. For this case, the benefits of using options are considerably greater.

# 5 Multiple Synergy Bidders

In this section, we consider market settings in which multiple synergy bidders are active simultaneously. Much of the experimental set-up and parameter choices are the same as described in the above sections, for the case of one for the single synergy bidder. The only difference is that now multiple synergy bidders may enter and leave the market at different times and they have different valuations for the combination of A and B.

We have to emphasize that the results from this section are still rather preliminary and are based on some restrictions on the reasoning capability of the synergy bidders in the market. Specifically, as in the single-bidder case, we assume the synergy bidders have some prior expectations about the closing prices in future auctions and compute their optimal strategy wrt. this expectation. In these results, this expectation is assumed the same for all synergy bidders, which is a reasonable choice in comparing their strategies. In a more realistic market, however, synergy bidders could be expected to be able to learn and adjust their expectations based on past interactions, as well as reason game-theoretically about the fact that another synergy bidder may present in the market at the same time. At this point, these more sophisticated forms of reasoning are left to future work.

As in the previous section all simulations of this section have reserve prices of 8 and local bidders following $\sim N(10, 2.5)$. The first two experiments also have two synergy bidders $syn_1$ and $syn_2$ with valuations for both goods of 21.5 and 22.5 respectively. The order the synergy bidders enter the market (and the number of auctions they can stay in) are given in Figs. 6 and 6, while results for all settings are shown in Fig. 7, respectively 9. In the following, we will discuss these in separate subsections.

## 5.1 Two Synergy Bidders Interacting Indirectly through the Exercise Price Level

In the setting examined here, the two synergy bidders each have $n_A = 3$ and $n_B = 3$, without the other agent participating in these auctions. An example of such an auction sequence is shown in Fig. 6.

This effect can be seen in Fig. 7, in which the seller maximizes her expected profit at $K = K_A = K_B = 2.4$. In this case $syn_2$ is better off, because without the presence of $syn_1$ she would be offered options with lower exercise prices. But $syn_1$ is worse off, because if she were alone in the market the seller would choose $K = 3.2$, which gives her a higher expected profit. Yet, due to $syn_2$, the seller sets

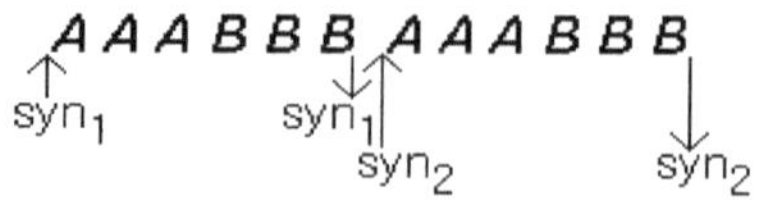


**Fig. 6.** An auction sequence for the case shown in Fig 7.

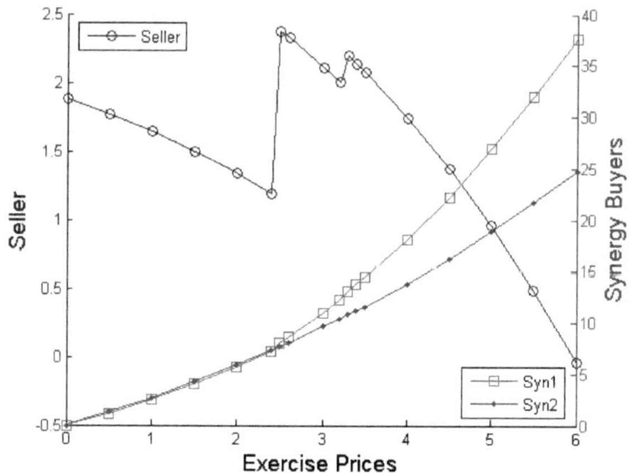


**Fig. 7.** Percentage increase in profits for a market with with 2 synergy bidders. There are 3 auctions for A and 3 for B, and for each one the bids from the competition formed by local bidders follows the distribution $N(10, 2.5)$. The valuations of the two synergy bidders for a bundle $\{A,B\}$ are 21.1 for $syn1$, respectively 22.5 for $syn2$. The order the agents enter the market is described by Fig. 6 below (so the two agents do not compete directly against each other in this setting). Notice that, in this case, the average profit of $syn2$ does not decrease with the entry of $syn1$ in the market.

$K = 2.4$. In this case, due to the seller's choice of exercise prices, one synergy bidder ($syn_1$) gains, while $syn_2$ loses.

## 5.2 Direct Synergy Bidder Competition in the Same Market

Next, we considered a setting in which synergy bidders compete directly for some of the goods. The entry points for such a setting are shown in Fig. 8, while simulation results are given in Fig. 9.

So, in this case, both synergy bidders lose from the presence of additional bidders. While one synergy bidder (i.e. $syn_2$) should benefit because she is offered better (higher) exercise prices than if she were alone in the market, this effect cannot immediately compensate the additional competition. Finally, we considered a larger scale simulation scenario involving multiple synergy bidders, which enter the market probabilistically at different time points.

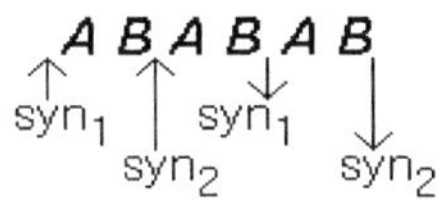


**Fig. 8.** An auction sequence for the case shown in Fig. 9.

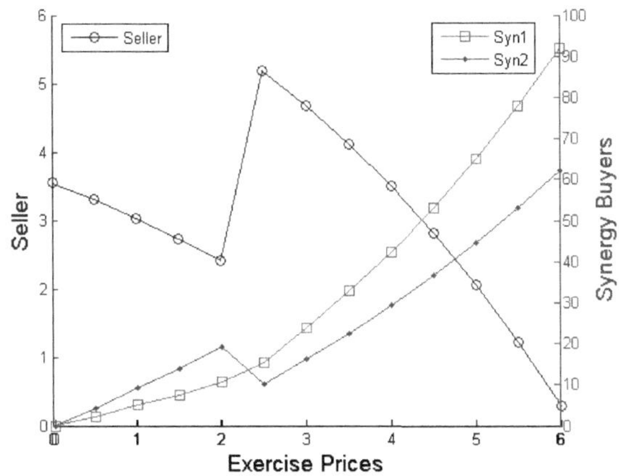


**Fig. 9.** Percentage increase in profits for a market with with 2 synergy bidders. The setting and valuations are the same as in Fig. 7 above. However, the order the agents enter the market is now described by Fig. 8 below (so the two agents *do* compete directly for the same goods). Notice that, in this case, the average profit of $syn2$ decreases due to the additional competition from $syn1$.

### 5.3 Simulation of Markets with Probabilistic Entry of Synergy Bidders

The experimental setup used for this case is as follows. Each sequence of auctions considered (forming one test case) has 10 items of each type (i.e. $n_A = 10$ and $n_B = 10$). What differs from previous settings is the random entry of synergy bidders. For each auction, there is a 25% chance that a synergy bidder will enter the market. If she does, then her valuation is drawn from a uniform distribution between 20 and 22 and she will stay in the market for exactly four auctions. To simplify matters, the auction sequence is fixed at first selling A, then B, then A etc. so that each synergy bidder will face exactly two auctions for an item of type A and two for an item of type B. However, the general result of this section also holds for a random auction sequence, since the basic effects remain the same.

As shown in Figure 10, the seller's profit now only has one maximum at 5, because initially each increase in exercise prices causes, with some probability, a synergy bidder to participate more often. So each point is a threshold and the profit graph smooths out over those many local maxima, corresponding to a steady increase (on average) of the expected profit. This result shows why it can be rational for the seller to have the same exercise prices for all goods of the same type

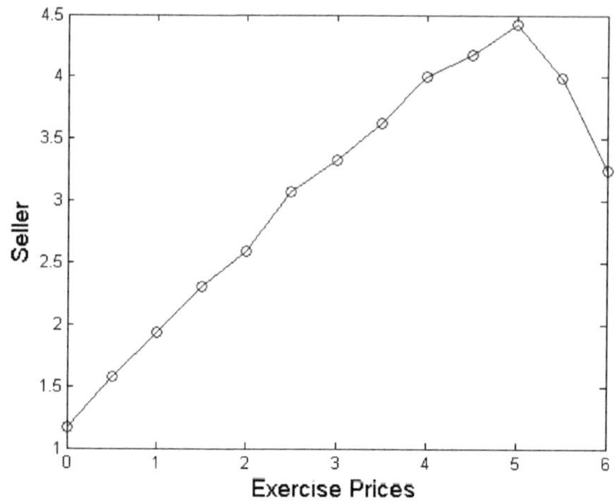


**Fig. 10.** Percentage increase in seller's profits in a larger experimental setting, with synergy bidders randomly entering the market.

(e.g. the same $K_A$). In a market with random entry of synergy bidders, the seller does not know which buyers are participating in any particular auction. Her optimal policy is to set her exercise prices which maximize her overall expected profit (here, $K = 5$).

## 6 Discussion and Further Work

This paper examined, from a decision-theoretic perspective, the use of priced options as a solution to the exposure problem in sequential auctions. We consider a model in which the seller is free to fix the exercise price for options on the goods she has to offer, and then sell these options through a first price auction.

For this setting, we derived analytically, for a market with a single synergy bidder, the expressions that provide the bounds on the option prices between which both synergy bidders and sellers have an incentive to use an option contract over direct auctions. Next, we performed an experimental analysis of several settings, where either one or multiple synergy bidders are active simultaneously in the market. We show that, if the exercise price is chosen correctly, selling items through priced options rather than direct sale can increase the expected profits of both parties.

The overall conclusion of our study is that the proposed priced options mechanism can considerably reduce the exposure problem that synergy bidders face when taking part in sequential auctions. Furthermore, and most important, *both* parties in the market have an incentive to prefer and use such a mechanism. We show that in many realistic market scenarios, sellers can fix the exercise prices at a level that both provides sufficient incentive for buyers to take part in the auctions, as well as cover their risk of remaining with the items unsold.

This work leaves several problems and aspects to be explored in further work. One such problem is that the presence of a fixed exercise price set by the seller can exclude some bidders from the market entirely (i.e. those bidders whose valuations are under the level of the exercise price). If the exercise price is set by the seller at a too high level, this may result in items remaining unsold, which would results in both a lower market allocation efficiency, as well as a potential loss to seller revenues. To deal with this problem, recent work by some of the authors of this paper [26] proposes a novel mechanism where *both* the options and exercise price are set dynamically by the market, based on the bids placed in each auction.

Another important, open issue left to further research is the use of more sophisticated reasoning abilities on the part of participating synergy bidders and sellers. For example, in a large market, synergy bidders could be expected to use learning strategies to adapt to changing market conditions, as well as the presence of other synergy bidders who want similar item combinations. However, the sellers of the items could also use learning to choose better levels of the exercise prices $K$ with which to sell the options for their goods. Other possible issues open to future research include: markets where bidders have asymmetric or imperfect information, more complex preferences over bundles and different attitudes to risk of the involved parties. Also, the applicability of distributed auctions and option techniques in modeling the allocation of user clicks in social web applications (see [24, 4, 19] for some examples) could be potentially investigated in further research.

To conclude, sequential auction bidding with complementary valuations is a problem that appears in many real-life settings, although no dominant strategies exist and bidders face a severe exposure problem. The main intuition of this work is that a simple options mechanism, where sellers auction options for their goods (with a pre-set exercise price), instead of the goods themselves can go a long way in solving the exposure problem, and can be beneficial to both sides in such a market.

## References


1. Boutilier, C., Goldszmidt, M., Sabata, B.: Sequential auctions for the allocation of resources with complementarities. In: Proc. of IJCAI 1999, pp. 523–527. Morgan Kaufmann, San Francisco (1999)
2. Gopal, R., Thompson, S., Alex Tung, Y., Whinston, A.B.: Managing risks in multiple online auctions: An Options approach. Decision Sciences 36(3) (2005)
3. Greenwald, A., Boyan, J.: Bidding under uncertainty: theory and experiments. In: AUAI 2004: Proceedings of the 20th conference on Uncertainty in artificial intelligence, Arlington, Virginia, United States, pp. 209–216. AUAI Press (2004)
4. Halpin, H., Robu, V., Shepherd, H.: The complex dynamics of collaborative tagging. In: Proceedings of the 16th Int. World Wide Web Conference (WWW 2007), pp. 211–220. ACM Press, New York (2007)
5. Hull, J.C.: Options, Futures, and Other Derivatives, 5th edn. Prentice-Hall, Englewood Cliffs (2003)
6. Ito, T., Klein, M.: A multi-issue negotiation protocol among competitive agents and its extension to a nonlinear utility negotiation protocol. In: AAMAS 2006: 5th Int. Conf. on Autonomous Agents and Multi-Agent Systems, pp. 435–437 (2006)

7. Ito, T., Zhang, M., Robu, V., Fatima, S., Matsuo, T.: Advances in Agent-Mediated, Complex Automated Negotiations. Springer Series on Computational Intelligence, vol. 233. Springer, Heidelberg (2009)
8. Ito, T., Zhang, M., Robu, V., Matsuo, T.: Proceedings of the 3rd Int. Workshop on Rational, Robust, Secure Negotiations. In: IEEE/ACM Int. Conf. on Intelligent Agent Technology - Workshops. IEEE Press, Los Alamitos (2007)
9. Jonker, C., Robu, V.: Automated multi-attribute negotiation with efficient use of incomplete preference information. In: 3rd Int. Conf. on Autonomous Agents & Multi Agent Systems (AAMAS), New York, pp. 1056–1063 (2004)
10. Jonker, C., Robu, V., Treur, J.: An agent architecture for multi-attribute negotiation using incomplete preference information. Journal of Autonomous Agents and Multi-Agent Systems (JAAMAS) 15, 221–225 (2007)
11. Jordan, P.R., Wellman, M.P.: Trading agent competition: Ad auctions (2009), `http://www.w3.org/DesignIssues/Fractal.html`
12. Juda, A., Parkes, D.: An options-based solution to the sequential auction problem. Artificial Intelligence (2009) (to appear)
13. Juda, A.I., Parkes, D.C.: An options-based method to solve the composability problem in sequential auctions. In: Agent-Mediated Electronic Commerce VI, pp. 44–58. Springer, Berlin (2006)
14. Juda, A.I., Parkes, D.C.: The sequential auction problem on ebay: An emperical analysis and a solution. In: Proc. of the 7th ACM Conf. on Electronic commerce, pp. 180–189. ACM Press, New York (June 2006)
15. Lai, G., Sycara, K., Li, C.: A decentralized model for automated multi-attribute negotiations with incomplete information and general utility functions. Multiagent Grid Systems 4(1), 45–65 (2008)
16. Lomuscio, A.R., Wooldridge, M., Jennings, N.R.: A classification scheme for negotiation in electronic commerce. Group Decision and Negotiation 12(1), 31–56 (2003)
17. Mous, L., Robu, V., Poutré, H.L.: Can priced options solve the exposure problem in sequential auctions? In: ACM SIGEcom Exchanges, vol. 7(2). ACM Press, New York (2008)
18. Mous, L., Robu, V., Poutré, H.L.: Using priced options to solve the exposure problem in sequential auctions. In: Proc. of Agent-Mediated Electronic Commerce (AMEC 2008). LNCS (LNBI). Springer, Heidelberg (2008)
19. Robu, V., Halpin, H., Shepherd, H.: Emergence of consensus and shared vocabularies in collaborative tagging systems. ACM Transactions on the Web 3(4), 1–34 (2009)
20. Robu, V., Poutré, H.L.: Designing bidding strategies in sequential auctions for risk averse agents. In: Rechnerstrukturen und Betriebsprogrammierung. LNBIP, vol. 13, pp. 76–89. Springer, Heidelberg (1974)
21. Robu, V., Noot, H., Poutré, H.L., van Schijndel, W.J.: An agent platform for auction-based allocation of loads in transportation logistics. In: Proc. 7th Int. Joint Conf. on Autonomous Agents and Multi Agent Systems (AAMAS 2008), Industry Track, pp. 3–10. IFAAMAS Press (2008)
22. Robu, V., Poutré, H.L.: Learning the structure of utility graphs used in negotiation through collaborative filtering. In: Proceedings of the Eighth Pacific Rim Workshop International on Multi-Agent Systems (PRIMA 2005), Kuala Lumpur, Malaysia (2005)
23. Robu, V., Poutré, H.L.: Retrieving the structure of utility graphs used in multi-item negotiation through collaborative filtering of aggregate buyer preferences. In: Proc. of the 2nd International Workshop on Rational, Robust and Secure Negotiations in Multi-Agent Systems (RRS 2006), Hakodate, Japan. Springer Series in Computational Intelligence, vol. 89 (2008)

24. Robu, V., Poutré, H.L., Bohte, S.: The complex dynamics of sponsored search markets. In: Cao, L., Gorodetsky, V., Liu, J., Weiss, G., Yu, P.S. (eds.) Agents and Data Mining Interaction. LNCS, vol. 5680, pp. 183–198. Springer, Heidelberg (2009)
25. Robu, V., Somefun, D.J.A., La Poutré, J.A.: Modeling complex multi-issue negotiations using utility graphs. In: Proc. of the 4th Int. Conf. on Autonomous Agents & Multi Agent Systems (AAMAS 2005), Utrecht, The Netherlands, pp. 280–287. ACM Press, New York (2005)
26. Robu, V., Vetsikas, I.A., Gerding, E.H., Jennings, N.R.: Flexibly priced options: A new mechanism for sequential auction markets with complementary goods. In: 9th Int. Conf. on Autonomous Agents and Multi-Agent Systems (AAMAS 2010) (2010)
27. Sandholm, T.: Algorithm for optimal winner determination in combinatorial auctions. Artificial Intelligence 135(1-2), 1–54 (2002)
28. Sandholm, T., Lesser, V.: Leveled commitment contracts and strategic breach. Games and Economic Behavior 35(1-2), 212–270 (2001)
29. Sandholm, T., Lesser, V.: Leveled-commitment contracting: a backtracking instrument for multiagent systems. AI Magazine 23(3), 89–100 (Fall 2002)
30. Hoen, P.J., Robu, V., Poutré, H.L.: Decommitment in a competitive multi-agent transportation setting. In: Software Agent-Based Applications, Platforms and Development Kits, Birkhauser. Springer, Heidelberg (2005)
31. van der Putten, S., Robu, V., Poutré, H.L., Jorritsma, A., Gal, M.: Automating supply chain negotiations using autonomous agents: a case study in transportation logistics. In: Proc. 5th Int. Joint Conf. on Autonomous Agents and Multi Agent Systems (AAMAS 2006), Industry Track, pp. 1506–1513. ACM Press, New York (2006)

# Desire-Based Negotiation in Electronic Marketplaces

Fenghui Ren and Minjie Zhang

School of Computer Science and Software Engineering,
University of Wollongong, Australia
{fr510,minjie}@uow.edu.au

**Summary.** Electronic Commerce has been a very significant commercial phenomenon in recent years and autonomous agents have made the advantage of e-marketplace to be more distinct. By comparison with traditional markets, the e-marketplace can bring more benefits to its participators. However, when an e-market environment becomes open and dynamic, existing agent negotiation models may expose some limitations in such a complex situation. Static negotiation strategies and encounter rules are difficulty to capture changes of market situations and omit potential impacts on negotiators' profits. For example, shoppers' benefits may be damaged when a market shifts away from the seller's market to the buyer's market. In this paper, we propose a novel desire-based negotiation model to solve such a challenge in open and dynamic e-marketplace through capturing changes of market situation and potential impacts on agents' reserved expectations. Experimental results illustrate the benefits and efficiency of the proposed model in complex e-market environments.

## 1 Introduction

Electronic Commerce (e-commerce) has been changing traditional methods of business in recent years and has become a very important commercial phenomenon. Nowadays, many businesses operate in e-marketplaces. By comparing with traditional markets, an e-market can effectively save participators' resources. For example, in e-marketplaces, merchants can save their budgets on business maintenance by avoiding physical shops and shop assistants. Also, shoppers do not need to visit shops in person which can save costs on traffic and time. Moreover, all participators can collect information about their concerned items and communicate with potential trading partners in a timely manner. Furthermore, autonomous agents have made the advantage of e-marketplaces to be more distinct. By employing autonomous agents, participators can even be free from information retrieval and bargaining, but they just 'tell' agents their expectations. Then the agents will negotiate with potential trading partners on behalf of their clients.

Negotiation among agents has been an important research area in agent and multi-agent systems for many years. The literature indicates great achievements from researchers. Faratin et al. [3] [17] presented several formal negotiation models among

T. Ito et al. (Eds.): Innovations in Agent-Based Complex Automated Negotiations, SCI 319, pp. 27–47.
springerlink.com 

autonomous agents and defined a number of strategies and tactics for different negotiation purposes in service oriented applications. Lai et al. [10] [11] proposed a non-biased mediator approach to help agents to achieve Pareto optimality and to overcome the difficulty of decisions due to incomplete information and lack of explicit utility functions. Fatima et al. [4] [5] [6] studied negotiation models in incomplete information settings in different negotiation scenarios and illustrated equilibrium solutions in different negotiation agendas and procedures. Besides these works on static negotiation environments, some works on complex negotiation environments have also been developed. Fatima et al. [7] proposed negotiation strategies to help agents to achieve approximately optimal outcomes in negotiations in order to improve computational efficiency. Their approach can be employed in dynamic negotiation environments with acceptable loss of equilibrium. Mason et al. [14] proposed price prediction strategies to help agents to estimate possible changes of markets and consequences from these changes. The authors also demonstrated that the proposed prediction strategies can help agents to improve their profits in dynamic markets. Kurbel et al. [9] introduced a model with fuzzy constraints on an e-job marketplace. They also proposed a negotiation protocol and negotiation strategies to address the challenge of multi-lateral negotiation in complex e-marketplaces. Furthermore, Li et al. [13] proposed confidence-based negotiation in complex agent environments. According to agents' confidence on negotiation partners, agents may apply different strategies and/or procedures. Agents can also update their profiles on partners in order to improve both processes and outcomes of their negotiations.

Although the above works have reached great achievements in solving some problems in agent negotiation, some challenges are still exist in e-marketplace negotiation. Most existing agent negotiation models requiring agents to predefine reserved expectations on negotiation outcomes before negotiations start, and agents' actions in encounter during the negotiation will be decided mostly by the reserved expectations. However, according to our studies on several models of e-marketplaces [1] [15] and e-marketplaces in the real world, we notice that in open and dynamic e-marketplaces, changes of a market situation may impact agents' reserved expectations on negotiation outcomes. In dynamic e-marketplaces, agents are not necessary to fix their reserved expectations in all situations. When market situations change, agent should modify their reserved expectation dynamically, during negotiation. We further noticed that if agents fix reserved expectations in dynamic e-marketplace, their profits might be damaged. That is because: (1) agents may have no clear idea about market situations, and the reserved expectation might be given by agents blindly; and (2) evaluation results based on fixed reserved expectations may not indicate items' real value in different market situations. For example, if an inexperienced house purchaser predefines his/her reserved offer without carefully investigation of the real estate market, it may lead the purchaser to two possible disadvantageous outcomes. (1) The house purchaser may undervalue properties' values in the market and does not accept any price higher than the reserved offer, so the potential purchaser may not find any satisfied property in the market and negotiations between all property sellers might fail. And (2) the house purchaser may overvalue properties' values in the market. Even though the purchaser can finally find a property in the market, his/her profit will be damaged as well. In order to solve such an issue in dynamic e-marketplaces, we propose a desire-based negotiation model to dynamically modify agents' reserved expectations during the negotiation based on changes of market situations and agents desires on accomplishment of a negotiation.

The rest of this paper is organized as follows. Section 2 proposes the desire-based negotiation model, which includes an offer evaluation approach, a counter-offer generation approach and a negotiation protocol. Section 3 illustrates experimental results of the proposed model in different market situations. Section 4 compares the proposed model with some related works, and Section 6 concludes the paper and outlines future work.

## 2 Desire-Based Negotiation Model

In this section, we introduce a desire-based negotiation model in order to help agents perform wise negotiations in complex e-marketplaces.

### 2.1 Principle

Through our studies, we notice that in the real world, although people can predefine reserved expectations in advance, in most cases it is not necessary for them to insist on their reserved expectations throughout the negotiation. For example, in a dynamic market, a hesitant buyer may look forward to gain more benefit when he/she notices that his/her initial expectation can be satisfied easily by most sellers. On the other hand, a 'rushing' buyer may accept an offer even if the offer is worse than his/her reserved expectation. However, most existing agent negotiation models do not take these situations into account. The motivation of this research is to introduce a desire-based negotiation model to help agents to make more wise actions in complex e-marketplaces.

In the desire-based negotiation model, agents do not need to predefine their reserved expectations, because the reserved expectations may be changed when the market changes. However, agents need to provide a figure to indicate their eagerness to reach the agreement in negotiations. And we make an assumption that agents' motivations on completion of the negotiation will not be changed during the negotiation. The reasons for such an assumption are based on two considerations. (1) The trading process in e-marketplace usually can be completed in a short time, and participators may not change their motivations and eagerness on completion of the trading in a short time. And (2) agents may modify their reserved expectations during the negotiation when the market situation changes in order to maximise their profits. If agents ensure that their profits will not be damaged when the market situation changes, they will not change their minds on completion of the trading. Therefore, by comparison with the reserved expectation, agent's *desire for trading* is more suitable to indicate agent's thought on the outcome of negotiations. In general, a rushing agent may have a high desire to complete the negotiation and a hesitant agent may have a low desire to complete the negotiation. During negotiation, agents may adjust reserved expectations based on their desires on completion of the negotiation when the market situation changes. Because agents may modify their reserved expectations in different market situations, agents may also adjust their standards to evaluate items' values and opponents' offers. So a same offer may be evaluated by agents differently in different market situations. For example, a buyer may accept a higher price for a car in a sellers' market, but the buyer will definitely reject the same offer in a buyers' market. That is because when the market situation changes, agents may have totally different evaluation results on opponents' offers. An advantageous offer may become disadvantageous, and vice versa. Therefore, in our desire-based negotiation model, agents will adjust their evaluations on offers and expectations on outcomes when market situation changes, and keep the balance between the profit of negotiation outcome and the completion of negotiation based on agents' *desire for trading*. Details of the desire-based negotiation model are introduced in the following subsections.

## 2.2 Offer Evaluation

In this subsection, we introduce an offer evaluation approach by considering both e-market situations and agent situations. It is proposed that the consideration on the situation of an e-market includes the number of consumers, the number of suppliers and the agent's role, while consideration on the situation of an agent includes the agent's subjective opinions on the e-marketplace and the negotiation.

### 2.2.1 Consideration of Market Situation

Before we introduce the offer evaluation approach, we will define some notations. Let tuple $< s, c, o_{ini}, d, \tau, \alpha, \beta, \lambda >$ be an indicator employed by agents during negotiation, where $s$ ($s > 0$) denotes the number of suppliers, $c$ ($c > 0$) denotes the number of consumers, $o_{ini}$ denotes the agent's initial offer, $d$ ($0 \leq d \leq 1$) denotes the agent's *desire for trading*. When $d = 0$, it indicates that the agent is not intent in completing the negotiation at all, and when $d = 1$, it indicates that the agent need to reach a negotiation agreement extremely. $\tau$ is the negotiation deadline. $\alpha$ denotes the agent's role in the negotiation, where $\alpha = -1$ for consumers and $\alpha = 1$ for suppliers. $\beta$ denotes the agent's attitude on markets' changes and $\lambda$ is the agent's bargaining strategy. Firstly, the relationship between supply and demand of an e-market at a certain moment is represented as follows:

$$\Phi(s, c, \alpha) = \frac{c - s}{c + s} \times \alpha \tag{1}$$

The value of Equation (1) is in the range of $[-1, 1]$, which indicate situations of the e-market environment for negotiators by considering negotiators' role in the negotiation. If $0 < \Phi \leq 1$, the e-market environment is in a beneficial state and the agent has an advantage in such an environment. If $-1 \leq \Phi < 0$, the environment is in an inferior state and the agent has a disadvantage in the environment. If $\Phi = 0$, the environment is in an equitable state and all agents play fairly in such an environment. Objectively, the Equation (1) represents the relationship between supply and demand in the negotiation environment at a certain moment. However, even for the same e-market situation, agents may also have their own considerations based on individual judgements. Therefore, we generate a graph (see Figure 1) to indicate the relationship between e-market environment and agents responses. In Figure 1, the $x$-axis represents situations of the e-market environment ($\Phi$), and the $y$-axis indicates negotiator responses. In general, when market situations shift away from a equitable state to a beneficial or an inferior state, an agent's responses will shift away from calmness to vehemence. In detail, it can be seen that agents may have three typical attitudes in response to changes of the market.

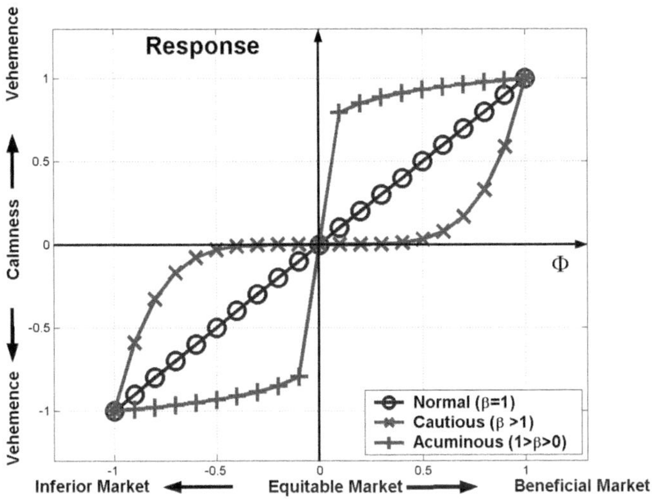


**Fig. 1.** Negotiators' responses to markets' situations

- Cautious ($\beta > 1$): when an environment's state shifts away from equitable to beneficial or inferior, negotiator's responses are very calm when changes of the environment are not significant. However, when changes in the environment are evident, negotiator responses will become more vehement.
- Acuminous ($1 > \beta > 0$): when an environment's status shifts away from equitable to beneficial or inferior, negotiators perform very sensitively even though the change in the environment is not very obvious. However, when an environment's status changes a lot, negotiators have to control their responses for some objective reasons (e.g. negotiators cannot make further concession anymore).
- Normal ($\beta = 1$): when an environment's state shifts away from equitable to beneficial or inferior, negotiator's responses are also shifted from calmness to vehemence reposefully.

Based on the above description, we generate the following mapping function from markets states to agents responses:

$$\Psi(s,c,\alpha,\beta) = \begin{cases} \Phi(s,c,\alpha)^{\beta}, & \Phi(s,c,\alpha) \geq 0 \\ -[-\Phi(s,c,\alpha)]^{\beta}, & \Phi(s,c,\alpha) < 0 \end{cases} \tag{2}$$

where $s$, $c$ and $\alpha$ are defined in Equation (1). The result of $\Psi$ indicates an agent's individual judgement about the e-market's situation. Different agents may have different judgements about the same e-market. When $\Psi > 1$, an agent estimates the e-market in a beneficial state, when $\Psi = 1$, an agent estimates the e-market in an equitable state, and when $\Psi < 1$, an agent estimates the e-market in an inferior state. However, because $\Psi$ only takes into account e-market situations, we also propose the following function to consider agents' individual situation in offers evaluation.

#### 2.2.2 Consideration of Negotiators Situation

Let $o_l$ denote an offer from an opponent and $o_{ini}$ denote negotiator's initial offer, then $o_l$ is evaluated by the negotiator as follows:

$$\Lambda(o_l, o_{ini}, \gamma) = \mathbf{th}\left(\frac{o_l - o_{ini}}{o_{ini}} \times \gamma\right) + 1 \tag{3}$$

where $\gamma = -1$ for issues which an agent prefers a lower value and $\gamma = 1$ for issues which an agent prefers a greater value, $\mathbf{th}(x)$ is defined as follow.

$$\mathbf{th}(x) = \frac{e^x - e^{-x}}{e^x + e^{-x}} \tag{4}$$

The result of Equation 3 ($\Lambda \in (0,2)$) indicates how Negotiator $P$'s initial offer is satisfied by the offer $o_l$. For example, if the negotiator plays as a consumer, when $o_l = o_{ini}$ then $\Lambda = 1$. It means that the consumer's expectation is satisfied. When $o_l > o_{ini}$ then $0 \leq \Lambda < 1$, it indicates that the consumer's expectation can only be achieved at a certain level. And when $o_l < o_{ini}$ then $1 < \Lambda < 2$, it implies that the consumer can gain more profit than he/she expected.

#### 2.2.3 Considerations of Both Markets and Negotiators Situation

Because Equation (3) only evaluates an offer based on the negotiators' initial offer but does not take the market situation into account, so evaluation results may not accord with the market. Therefore, we define the offer evaluation function by considering both market and negotiator situations as follows:

$$\Theta(o_l, s, c, o_{ini}, \alpha, \beta, \gamma) = \frac{\Lambda(o_l, o_{ini}, \gamma)}{\Psi(s,c,\alpha,\beta) + 1} \tag{5}$$

Where the result of Equation 5 ($\Theta \in (0,1)$) indicates the negotiator's utility by accepting the offer $o_l$ in a certain market. If it is an equitable market ($\Psi = 0$ and

$\Theta = \Lambda$), then the offer $o_l$ is evaluated unbiasedly. If it is in a beneficial market ($0 < \Psi < 1$ and $\Theta < \Lambda$), then the offer $o_l$ is undervalued. And if it is in an inferior market ($-1 < \Psi < 0$ and $\Theta > \Lambda$), then the offer $o_l$ is overvalued.

For example, if a potential car purchaser's initial offer is \$6000 and the seller's reserved price is \$6500. Without consideration of the market situation, the buyer's evaluation result on the offer \$6500 is $\Lambda = 0.92$. So the buyer is 92% satisfying with the seller's offer. However, when taking the market situation into account, results might be different. If the market is a buyers' market (i.e. 5 buyers and 10 sellers, then $\Psi = 1.4$), the buyer's satisfaction on the offer \$6500 will decrease to 66% and the buyer may reject the offer. That is because in the buyers' market, a buyer has opportunities to make greater profits. On the other hand, if the market is a sellers' market (i.e. 10 buyers and 5 sellers, then $\Psi = 0.72$), the buyer's satisfaction on the offer \$6500 will increase to 128%. It indicates that the buyer is very happy on the offer \$6500 in a disadvantageous market and may accept the offer. During negotiation, desire-based agents will make decisions on their actions based on the result of Equation (5) and agents' *desire for trading* $d$ (see Section 2.4 for details). For editing reasons, we simplify the expression of Equation (5) to $\Theta(o_l)$ in the follows.

### 2.3 Counter-Offer Generation

In the last subsection, we introduce the approach to evaluate opponent's offers by consideration both the market and negotiator situations. In this subsection, we introduce the counter-offer generation approach. The counter-offer generation approach also takes both the market and negotiator situations into account. Before we introduce this approach, we define some notations.

Let set $\mathbf{O_t}$ denote all offers that an agent received from its opponents in round $t$, $o_t^b$ denote the 'best' offer in $\mathbf{O_t}$ (i.e. the offer brings the highest profit to the agent), $o_t^w$ denote the 'worst' offer in $\mathbf{O_t}$ (i.e. the offer brings the lowest profit to the agent), $o_t^m$ denote the average of $\mathbf{O_t}$ ($o_t^m = \frac{1}{N}\sum_{i=1}^{N} o_t^i$ and $N$ is the size of set $\mathbf{O_t}$), $o_{t'}^b$ denote the estimated best offer in the next round $t'$, $co_t$ denotes the agent's last counter-offer, and $co_{t'}$ denote the agent's counter-offer for the next round. Then if an agent plays as a buyer, one possible situation of the counter-offer generation procedure in round-$t$ is illustrated in Figure 2.

In Figure 2, the $x$-axis stands for prices and the $y$-axis stands for the occurrence density on each price. The solid curve indicates the distribution of set $\mathbf{O_t}$ in round-$t$, which may differ from case to case, and the dotted line is the estimated distribution of set $\mathbf{O'_t}$ in the next round. We make the assumption that the shape of the distribution curve of set $\mathbf{O'_t}$ is similar to $\mathbf{O_t}$'s, but just the range is changed. Because the agent plays as a buyer, so the market represented in Figure 2 is a beneficial market. In a beneficial market, for buyers, $\mathbf{O'_t}$ is estimated to be smaller than $\mathbf{O_t}$ on average. The distance between the current counter-offer $co_t$ and the estimated 'best' offer $o_{t'}^b$ in the next round is the bargaining area. The new counter-offer $co_{t'}$ is generated within this area according to the agent's negotiation strategy and remaining rounds as follows.

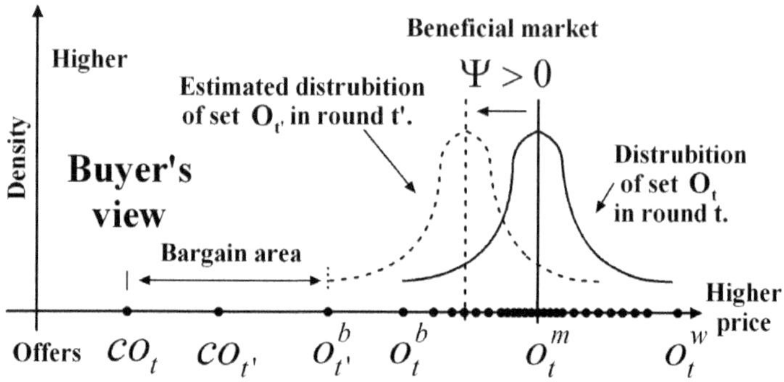


**Fig. 2.** Counter-offer generation

Firstly, we estimate the 'best' offer $o^b_{t'}$ in the next round $t'$ by considering both the distribution of $\mathbf{O_t}$ and the market situations as follows:

$$o^b_{t'} = o^b_t + \Psi(s, c, \alpha, \beta) \times \sqrt{D(\mathbf{O_t})} \times \gamma \tag{6}$$

$$D(\mathbf{O_t}) = \sum_{i=1}^{N} (o_{t,i} - E(\mathbf{O_t}))^2 p_i \tag{7}$$

where $D(\mathbf{O_t})$ indicates the variance of $\mathbf{O_t}$, $\gamma = -1$ for issues which an agent prefers a lower value and $\gamma = 1$ for issues which an agent prefers a greater value, $E(\mathbf{O_t})$ indicates the mathematical expectation of $\mathbf{O_t}$, $p_i$ indicates the distribution of $o_{t,i}$ and $\Psi(s, c, \alpha, \beta)$ indicates the agent's response to the market situation. Usually, when the distribution of $\mathbf{O_t}$ is a Gaussian distribution, then $E(\mathbf{O_t}) = o^m_t$, $p_i = \frac{1}{N}$ and Equation 7 is specified as:

$$D(\mathbf{O_t}) = \frac{\sum_{i=1}^{N} (o_{t,i} - o^m_t)^2}{N} \tag{8}$$

Then the counter-offer $co_{t'}$ for the following negotiation round is generated as follows:

$$co_{t'} = \begin{cases} o_{ini}, & when\, t = 0, \\ co_t + (o^b_{t'} - co_t) \times (\frac{t}{\tau})^{\lambda}, & when\, 0 < t \leq \tau. \end{cases} \tag{9}$$

where $o_{ini}$ is the agent's initial offer, $co_t$ is the agent's last counter-offer, $o^b_{t'}$ is the estimated 'best' offer in the next round, and we simply adopt parameter $\lambda$ in Faratin et al.'s model [3] to represent the negotiator's bargaining strategies.

In Figure 3, it can be seen that when the market becomes very beneficial to the buyer agent, it is possible that $o^b_{t'} < co_t$ and $co_{t'} < co_t$. So in the desire-based negotiation model, we propose a decommitment mechanism which allows agents to reject previous counter-offers if the counter-offer is not formally accepted by any opponents. The reason that we propose such a mechanism is because in the

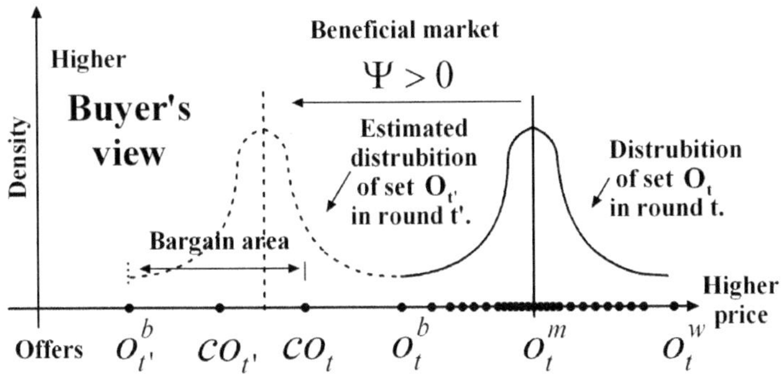


**Fig. 3.** Counter-offer generation

desire-based negotiation model, both the offer evaluation approach and counter-offer generation approach are impacted by market situations. So when the market situation changes, agents may change their considerations on both offer evaluation and counter-offer generation as well in order to gain more profits. For example, buyers may generate disadvantageous counter-offers when the market is inferior. However, when buyers notice that the market may become beneficial and if previous counter-offers are not accepted by any seller, buyers can reject the previous disadvantageous counter-offers and re-generate advantageous counter-offers in order to enlarge their profits. On the other hand, if sellers notice that the market may become inferior for them in advance, they may accept buyers' current offers in order to avoid losses in the future.

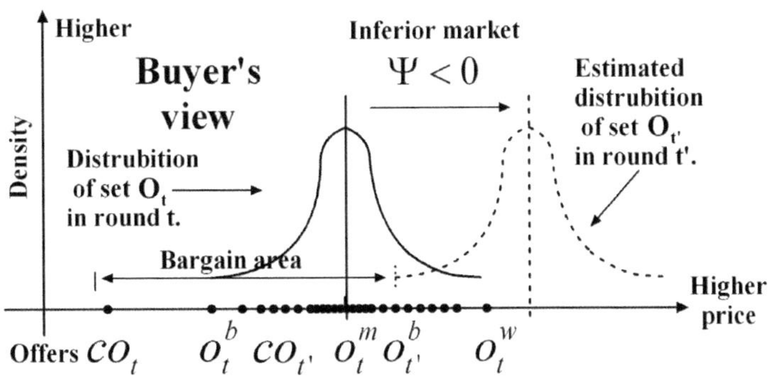


**Fig. 4.** Counter-offer generation

Also, markets may become inferior for buyers. In Figure 4, it can be seen that when a market is inferior for buyers, the estimated 'best' offer for the following round is worse than the 'best' offer in the round $t$, i.e. $o^b_{t'} > o^b_t$. During negotiations, if the new counter-offer in the round $t'$ can bring more profits to the agent than the 'best' offer in the current round $t$, i.e. $\Theta(co_{t'}) > \Theta(o^b_t)$, the negotiator will keep on bargaining with opponents and send out the new counter-offer $co_{t'}$. However, if the new counter-offer is worse than the 'best' offer from opponents, i.e. $\Theta(co_{t'}) < \Theta(o^b_t)$, (see the case shown in Figure 4), the agent will not send the new counter-offer $co_{t'}$, but make its final decision about the negotiation based on the comparison between the 'best' offer ($o^b_t$) from opponents and the agent's *desire for trading* ($d$). The detailed encounter rule of the desire-based negotiation model is introduced in the following subsection.

## 2.4 Negotiation Protocol

Since both the offer generation approach and the counter-offer evaluation approach have some differences from existing negotiation models [3] [8] [17], we propose a negotiation protocol for our desire-based negotiation model based on Rubinstein's alternating offers protocol [3] as follows.

Step 1: The agent assigns negotiation parameters, i.e., initial offer ($o_{ini}$), *desire for trading* ($d$), negotiation deadline ($\tau$), role in negotiation ($\alpha$), attitude on markets changes ($\beta$) and bargaining strategy ($\lambda$). The number of consumers ($c$) and suppliers ($s$) can be obtained from the environment directly. Also the agent initializes $t$ to 0 and $co_t$ to $o_{ini}$.

Step 2: The agent broadcasts $co_t$ to all opponents and waits for responses.

Step 3: Once the agent gets responses, if any opponent accepts $co_t$, the negotiation is completed. Otherwise, if $t > \tau$, the procedure goes to Step 4; and if $t \leq \tau$, the procedure goes to Step 5.

Step 4: Because the agent does not have time for further bargaining, it has to make a final decision on the 'best' offer $o^b_t$ in the last round. If $\Theta(o^b_t)$[1] $\geq 1-d$, the agent will accept $o^b_t$ and the negotiation is completed. Otherwise, the negotiation fails.

Step 5: Because the agent still has time for further bargaining, so the agent will generate a new counter-offer $co_{t'}$ for the next round. If $\max(\Theta(o^b_t), \Theta(co_{t'}), 1-d) = \Theta(o^b_t)$, the offer $o^b_t$ will be accepted by the agent and the negotiation is completed. If $\max(\Theta(o^b_t), \Theta(co_{t'}), 1-d) = 1-d$, the agent will leave off the procedure and the negotiation fails. If $\max(\Theta(o^b_t), \Theta(co_{t'}), 1-d) = \Theta(co_{t'})$, the procedure goes to Step 6.

Step 6: The agent updates $t$ to $t'$, $co_t$ to $co_{t'}$ and parameters $c$, $s$ according to the current market situation, then the procedure goes back to Step 2.

Based on the above procedure, the negotiator's action in round $t$ is defined as follows:

[1] Simplification of Equation (5).

$$\Omega(t) = \begin{cases} \textbf{Quit}, t \geq \tau \wedge \Theta(o_t^b) < 1 - d \textbf{ or} \\ t < \tau \wedge \max(\Theta(o_t^b), \Theta(co_{t'}), 1 - d) = 1 - d, \\ \textbf{Accept } o_t^b, t \geq \tau \wedge \Theta(o_t^b) \geq 1 - d \textbf{ or} \\ t < \tau \wedge \max(\Theta(o_t^b), \Theta(co_{t'}), 1 - d) = \Theta(o_t^b), \\ \textbf{Offer } co_{t'}, t < \tau \wedge \max(\Theta(o_t^b), \Theta(co_{t'}), 1 - d) = \Theta(co_{t'}). \end{cases} \quad (10)$$

# 3 Experiments

In this section, we illustrate our experimental results on the proposed desire-based negotiation model and compare our model with the NDF model [3]. Subsection 3.1 introduces the experimental setup. Subsection 3.2 demonstrates the experimental results. In Subsection 3.3, we analyze the experimental results and present further discussion on the proposed model.

## 3.1 Experimental Setup

In order to mimic situations of an e-marketplace, we employed five agents (two consumers and three suppliers). Both consumers want to purchase a monitor and all suppliers have a monitor to sell in different prices. One consumer (Agent $c2$) and all suppliers (Agents $s1$, $s2$ and $s3$) employ the NDF negotiation strategy. Their negotiation parameters are listed in Table 1. In order to simplify the experiment, deadlines for all agents are set to the $10^{th}$ round ($\tau = 10$) and all bargaining strategies are linear ($\lambda = 1$). The consumer (Agent $c1$) employs the proposed desire-based negotiation model and its negotiation parameters are $o_{ini} = \$100$, $\tau = 10$ and $d = \alpha = \beta = \lambda = 1$. The reason of setting the parameter $d$ to 1 is to ensure that agents will not leave off the negotiation in midway, so we can inspect all counter-offers generated by Agent $c1$. During the negotiation, all NDF agents employ the NDF negotiation protocol and Agent $c1$ employs the proposed negotiation protocol. All agents will keep their parameters in private and secure their counter-offers from competitors. Experiments are performed separately according to markets situations, i.e. a inferior market, an equitable market and a beneficial market (from consumers' view).

**Table 1.** Agents use NDF strategy

| Agent | $o_{ini}$ | $o_{res}$ | $\tau$ | $\lambda$ | $\alpha$ |
|---|---|---|---|---|---|
| c2 | \$100 | \$200 | 10 | 1 | $-1$ |
| s1 | \$200 | \$100 | 10 | 1 | 1 |
| s2 | \$300 | \$150 | 10 | 1 | 1 |
| s3 | \$250 | \$150 | 10 | 1 | 1 |

## 3.2 Experimental Results

In this subsection, we illustrate the experimental results on desire-based negotiation in an inferior market, an equitable market and a beneficial market (from the consumers' view point), respectively.

### 3.2.1 Inferior Market

In the experiment, we mimic the inferior market by involving two consumers and one supplier in the negotiation. In this scenario, consumers face competition. In order to win the competition, each consumer has to defeat its competitors. Experimental results between the two consumers ($c1$ and $c2$) and the one supplier ($s1$ or $s2$ or $s3$) are illustrated in Figures 5 through 7, respectively. It can be seen that in the inferior market, in order to defeat $c2$, $c1$'s offers have higher prices than $c2$'s in all negotiation rounds. $c1$ finally won all negotiations with different sellers, i.e. \$158.25 with $s1$, \$227.43 with $s2$, and \$199.64 with $s3$. The explanation about such results are (1) $c1$ noticed that the current market was inferior for itself and gives more concessions to sellers in order to get the item, and (2) because $c1$'s desire for trading was set to 1, it would accept any price finally in order to ensure the item can be gained. Of course, if we decrease the value of $c1$'s *desire for trading*, $c1$ might leave off the negotiation when it notices that the market is too bad to reach any agreement.

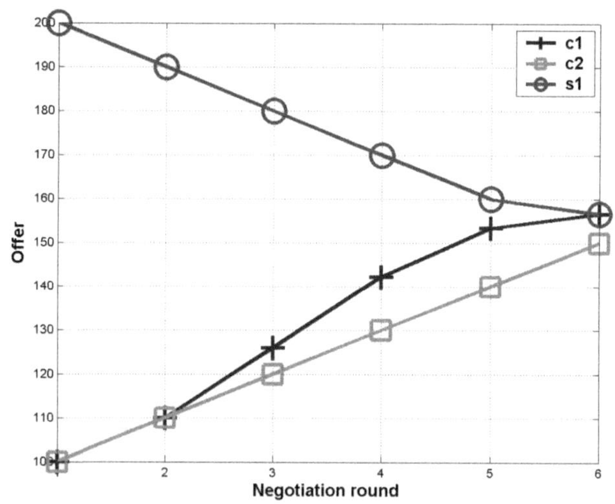


**Fig. 5.** Negotiations in a inferior market (c1, c2 vs. s1).

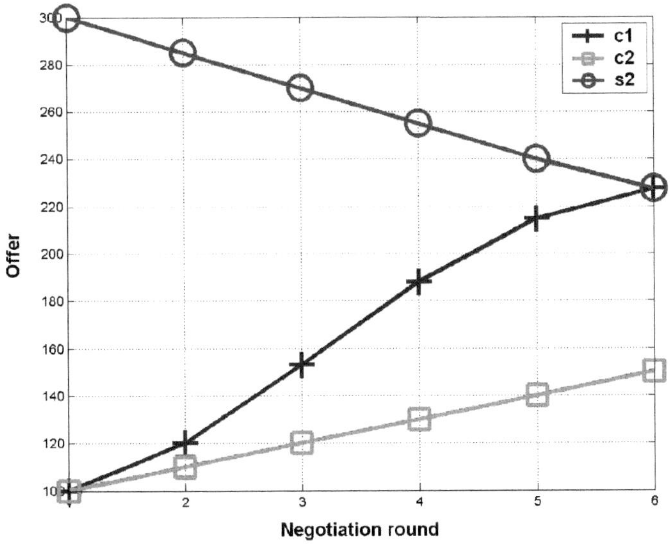


**Fig. 6.** Negotiations in a inferior market (c1, c2 vs. s2).

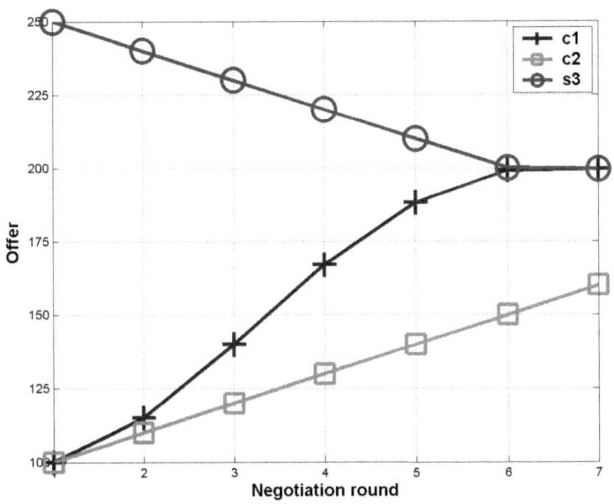


**Fig. 7.** Negotiations in a inferior market (c1, c2 vs. s3).

### 3.2.2 Equitable Market

We set both the number of consumer's and supplier's to two in order to mimic an equitable market. The experimental result is displayed in Figure 8. In the equitable market, $c1$ noticed that competitions are not so serious as in the inferior market, so in order to enlarge its profit, it would not generate offers in a very high price as it did in the inferior market, but just an offer slightly higher than $c2$'s price. It can be seen that $c1$ made a deal with $s1$ firstly in $\$156.66$, while $c2$ made a deal with $s2$ finally in $\$180.0$. Therefore, by comparison with $c2$, $c1$ gained more profit by adopting our proposed negotiation model in an equitable market. Also, $c1$ decreased its cost by comparison with its best agreement ($\$158.25$) in the inferior market.

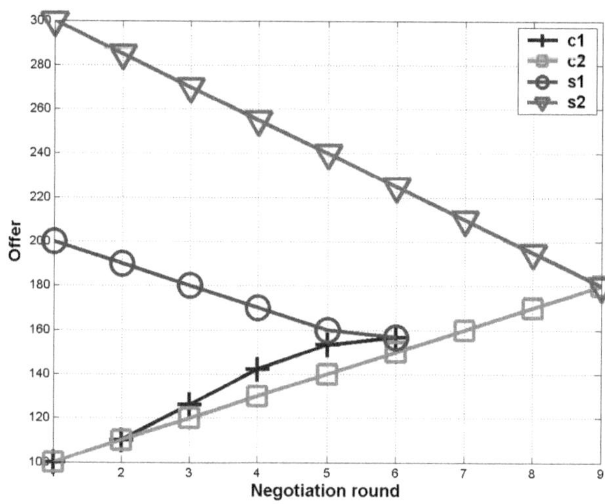


**Fig. 8.** Negotiations in a equitable market.

### 3.2.3 Beneficial Market

Experimental results in a beneficial market are displayed in Figures 9 through 13. Firstly, we did not put any competition pressures on consumers, so each consumer negotiated with two suppliers individually. It can be seen in Figure 9, when $c1$ negotiated with two suppliers ($s1$ and $s2$), it made a deal with $s1$ in $\$140.45$. Comparison to that, $c2$ made a deal with $s1$ in $\$150$ (see Figure 10). Obviously, $c1$ gained more profit than $c2$ from the same negotiation. Furthermore, when $s3$ entered into the negotiation, because the market became much advantageous for consumers, so $c1$ made a deal in a much lower price with $s1$ in $\$110$ (see Figure 11). However, $c2$ could not enlarge its profits even though the market became better and still made a deal with $s1$ in $\$150$ (see Figure 12). During the negotiation with multiple suppliers, it can be seen that $c1$ employed the decommitment mechanism we proposed in

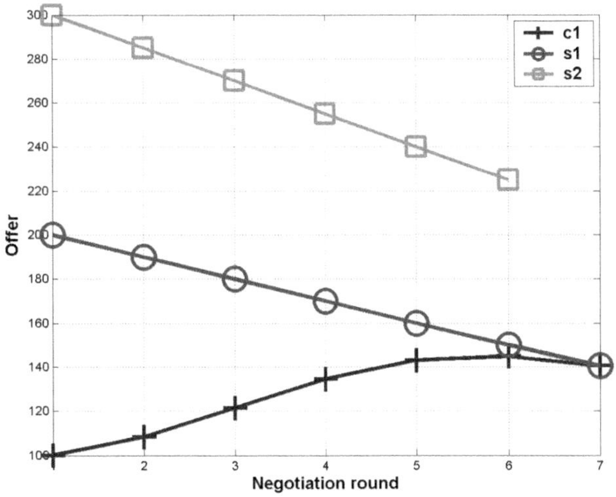


**Fig. 9.** Negotiations in a beneficial market (c1 vs. s1, s2).

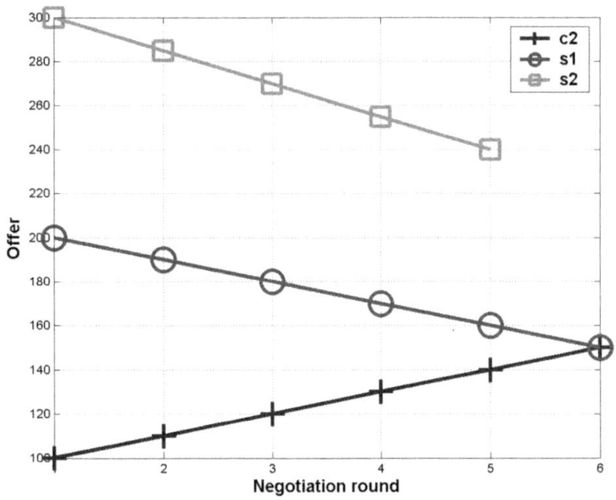


**Fig. 10.** Negotiations in a beneficial market (c2 vs. s1, s2).

Subsection 2.3. $c1$ resigned counter-offers rejected by all suppliers in the previous negotiation round, and re-generated counter-offers in a lower price when $c1$ notices the market is in beneficial state. By performing such a behavior, $c1$ successfully enlarge its profits in the beneficial market.

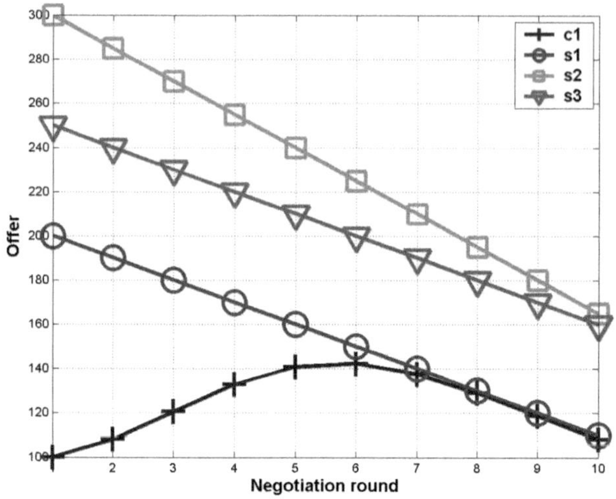


**Fig. 11.** Negotiations in a beneficial market (c1 vs. s1, s2, s3).

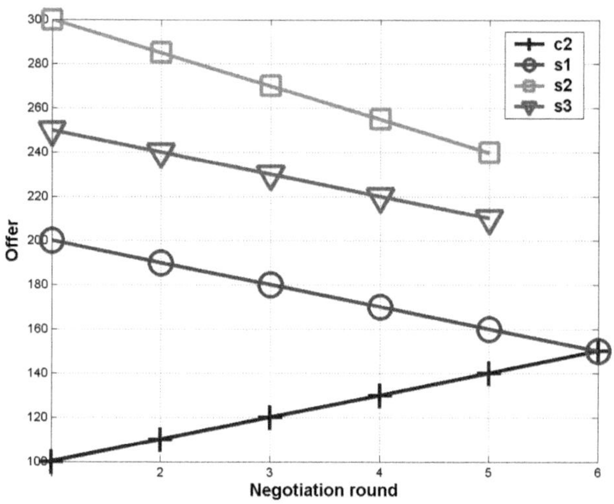


**Fig. 12.** Negotiations in a beneficial market (c2 vs. s1, s2, s3).

In Figures 13, we illustrate another situation in the beneficial market. In this case, $c1$ will face competition from another consumer $c2$. $c1$ firstly made a deal with $s1$ in $\$150.84$, which was higher than the price ($\$110$) in the beneficial market without any competition (see Figure 11), but was lower than the price ($\$156.66$) in the equitable market (see Figure 8) and the best price ($\$158.25$) in the inferior market (see Figure 5). Therefore, it can be confirmed that $c1$ has the ability to adjust its

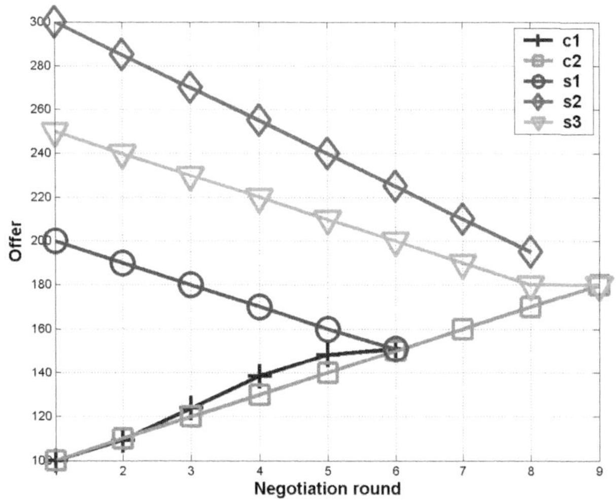


**Fig. 13.** Negotiations in a beneficial market.

negotiation behavior to balance between profit and success during the negotiation when markets situation change. Furthermore, by comparison with the cases displayed in Figure 9 and 11, it can be seen when $c1$ had the same increment on both numbers of competitors and partners, $c1$ lost its profit. Therefore, we may infer that competitions impact agents' profits more than opportunities. However, we will not expand such a discussion in this paper but leave it to future works.

### 3.3 Discussions

In the previous subsection, we illustrated experimental results in different market situations. It can be seen when a market changes, a desire-based agent also changes its behaviors during negotiations. Furthermore, even for the same market situation, an agent's decisions may also be different when they have different desires on trading. Therefore, both market situations and agent desires will impact negotiation results. In this subsection, we discuss how these two factors affect agent behaviors in negotiations.

In Figure 14, we illustrate a model to demonstrate how market situations and agent desires for trading impact agents behaviors in negotiations. The $x$-axis denotes the markets situation (refer to Equation (1)), the $y$-axis denotes agents' *desire for trading*, and the $z$-axis denotes agents satisfactions on offers (refer to Equation (3)). Then by setting both negotiation parameters $\beta$ and $\lambda$ to 1, a trading surface for the desire-based negotiation model can be formulated as follows:

$$\Gamma(\Phi, d) = \begin{cases} (1+\Phi) * (1-d), \ \textbf{when} \ -1 \leq \Phi \leq 0, \\ (\Phi - 1) * d + 1, \ \textbf{when} \ 0 < \Phi \leq 1. \end{cases} \tag{11}$$

where $d \in [0, 1]$ and $\Phi \in [-1, 1]$.

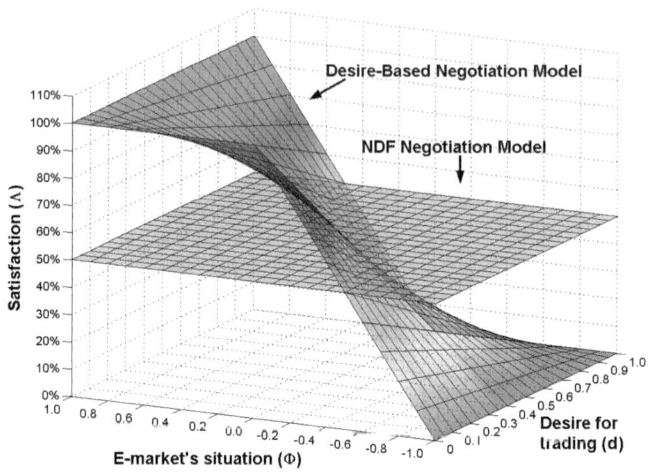


**Fig. 14.** Trading surface of negotiation models.

The trading surface defines a set of thresholds on agents profits. During the negotiation, agents will accept offers above or on the surface, but reject offers below the surface. For the trading surface of desire-based negotiation, in an extreme case, when $\Phi = -1$ or $d = 1$ then the threshold is $\Gamma(\Phi, d) = 0$, so the agent will accept the 'best' offer from its opponents finally in order to make the deal. That is because when $\Phi = -1$, the market is extremely disadvantageous for the agent, so any offer will be considered as a 'good' offer based on the market situation; and when $d = 1$, the agent needs to complete the negotiation extremely, so the agent will accept the 'best' offer from its opponents finally. In another extreme case, when $\Phi = 1$ or $d = 0$ then the threshold is $\Gamma(\Phi, d) = 1$, so the agent will reject any offer which cannot satisfy itself by $100\%$. That is because when $\Phi = 1$, the market is extremely advantageous for the agent, so any offer below the agent's initial offer will be considered as a 'bad' offer; and when $d = 0$, the agent's motivation for completing the negotiation is very low, so any offer lower than $100\%$ satisfaction definitely can not touch the agent. In a normal case, such as $\Phi = 0$ and $d = 0.5$ (i.e. an equitable market and the agent hesitates about trading), the agent will not accept any offer which cannot meet its satisfaction on $50\%$.

Also, we display another trading surface for the NDF model. Comparison with the desire-based model, the NDF model's trading surface is just a plane surface. That means the agent in the NDF model will fix its thresholds in all situations into a constant, and does not consider changes of markets and agents desires for trading. It can be seen in Figure 14 that an instance of the trading surface for the NDF model ($\Lambda = 50\%$) is partially below the trading surface of desire-based model and partially above the desire-based model's. For agents in the NDF model, they will accept all offers on this surface. However, for agents in the desire-based model, situations are more complex. When $\Phi > 0$ (beneficial market) and $d < 0.5$ (i.e. agents do not really want to make a deal), agents in the desire-based model will not accept offers which locate on the NDF model's surface. On the other hand, when $\Phi \leq 0$ (inferior or equitable market) and $d > 0.5$ (i.e. agents want to make a deal), the agent in the desire-based model will accept offers which locate on the NDF model's surface. Therefore, we can conclude that in the desire-based negotiation model, agents do not evaluate offers independently, but relatively by considering market situations and agent desires for trading, and the desire-based negotiation model is more applicable in complex e-marketplaces.

## 4 Related Works

Some related works also take into account agent negotiation in complex environments. This section discusses differences between these related works and our model.

Sycara et.al. [12] proposed a model for bilateral negotiation by considering uncertain and dynamic outside options. It is argued that outside options can impact agent negotiation strategies. According to the complexity of outside options in negotiation, negotiations are further divided into three levels, which are *single-thread negotiation*, *synchronized multi-thread negotiation* and *dynamic multi-thread negotiation*. *Single-thread negotiation* is only processed between two agents without outside options. *Synchronized multi-thread negotiation* is based on the *single-thread negotiation* model, and also considers concurrently existing outside options. *Dynamic multi-thread negotiation* is expanded from *synchronized multi-thread negotiation* by considering uncertain outside options which may occur dynamically in the future. Sycara's model gives a very novel classification and description on general negotiation. The desire-based negotiation model proposed in this paper focuses on e-market places and its changes, and belongs to *dynamic multi-thread negotiation*.

Dasgupta and Hashimoto [2] proposed an approach to address the problem of dynamic pricing in a competitive online economy where a product is differentiated by buyers and sellers on multi-issue. Agents may have incomplete knowledge of the negotiation parameters. A seller employs a collaborative filtering algorithm to determine a temporary consumer's purchase preferences and a dynamic pricing algorithm to determine a competitive price for the product. Therefore, the price prediction approach gives a solution about the bidding strategy in complex negotiation environments. However, their approach only pays attention to sellers without the consideration of the situation of buyers. Our desire-based negotiation model is suitable for adoption by both sellers and buyers.

Ren et.al. [16] proposed a market-driven model to help agents to make concessions in negotiation. Four concession factors, namely *trading opportunity*, *trading competition*, *trading time and strategy* and *eagerness*, are introduced to represent both market and agent situations. Each concession factor impacts an agent's concession from a certain consideration. All concession factors are updated by the agent according to the market's dynamic situation. But agents' judgements on offers and expectations on negotiation outcomes are still fixed. In this paper, we model markets by considering both market situations and agent desires. During negotiations, agents make concessions based on both objective and subjective considerations in the negotiation.

By comparison with the above related works, the proposed desire-based negotiation model has the following merits. It models negotiations in e-marketplaces by considering (1) both objective situations of markets and subjective desires of agents, (2) both concurrent and future possible situations of e-marketplaces, and (3) both agents' individual profit and trade-offs of whole e-market places.

## 5 Conclusion and Future Works

In this paper, we proposed a desire-based negotiation model to help agents to make more wise decisions in e-marketplaces by considering both market situations and agent desires for trading. In our model, the offer evaluation approach and counter-offer generation approach take both objective and subjective considerations into account. Offers from opponents are evaluated relatively by considering markets situations and counter-offers are generated wisely by expecting possible changes of the market in the future. Also, a negotiation protocol was proposed to define the negotiation procedure in e-marketplaces. Based on experimental results, we further put forward the concept of 'trading surface' and discovered that the trading surface of the desire-based negotiation model is more applicable than the NDF model's in complex e-marketplaces. Future works of this research will focus on multi-issue negotiations and analysis on competitions and opportunities in e-marketplaces.

## References

1. Das, S.: The Effects of Market-Making on Price Dynamics. In: 7th Int. Conf. on Autonomous Agents and Multiagent Systems (AAMAS 2008), pp. 887–894 (2008)
2. Dasgupta, P., Hashimoto, Y.: Multi-Attribute Dynamic Pricing for Online Markets Using Intelligent Agents. In: 3rd Int. Conf. on Autonomous Agents and Multiagent Systems (AAMAS 2004), pp. 277–284 (2004)
3. Faratin, P., Sierra, C., Jennings, N.: Negotiation Decision Functions for Autonomous Agents. Journal of Robotics and Autonomous Systems 24(3-4), 159–182 (1998)
4. Fatima, S., Wooldridge, M., Jennings, N.: Multi-Issue Negotiation Under Time Constraints. In: 1st Int. Conf. on Autonomous Agents and Multi-Agent Systems (AAMAS 2002), pp. 143–150 (2002)
5. Fatima, S., Wooldridge, M., Jennings, N.: An Agenda-Based Framework for Multi-Issue Negotiation. Artificial Intelligence 152(1), 1–45 (2004)
6. Fatima, S., Wooldridge, M., Jennings, N.: Optimal Negotiation of Multiple Issues in Incomplete Information Settings. In: 3rd Int. Conf. on Autonomous Agents and Multiagent Systems (AAMAS 2004), pp. 1080–1087 (2004)
7. Fatima, S., Wooldridge, M., Jennings, N.: Approximate and Online Multi-Issue Negotiation. In: 6th Int. Conf. on Autonomous Agents and Multi-Agent Systems (AAMAS 2007), pp. 947–954 (2007)

8. Jennings, N., Faratin, P., Lomuscio, A., Parsons, S., Sierra, C., Wooldridge, M.: Automated Negotiation: Prospects, Methods, and Challenges. Int. J. Group Decision Negotiation 10(2), 199–215 (2001)
9. Kurbel, K., Loutchko, I.: A Model for Multi-Lateral Negotiations on An Agent-Based Job Marketplace. Electronic Commerce Research and Applications 4(3), 187–203 (2005)
10. Lai, G., Li, C., Sycara, K.: A General Model for Pareto Optimall Multi-Attribute Negotiations. In: 2nd Int. Workshop on Rational, Robust, and Secure Negotiations in Multi-Agent Systems (RRS 2006), pp. 55–76 (2006)
11. Lai, G., Sycara, K., Li, C.: A Pareto Optimal Model for Automated Multi-attribute Negotiations. In: 6th Int. Conf. on Autonomous Agents and Multi-Agent Systems (AAMAS 2007), pp. 1040–1042 (2007)
12. Li, C., Sycara, K., Giampapa, J.: Dynamic Outside Options in Alternating-Offers Negotiations. In: 38th Annual Hawaii Int. Conf. on System Sciences (HICSS 2005), pp. 1–10 (2005)
13. Li, X., Soh, L.: Adaptive, Confidence-Based Strategic Negotiations in Complex Multiagent Environments. In: 9th Int. Florida Artificial Intelligence Research Society Conf., pp. 80–85 (2006)
14. MacKie-Mason, J., Osepayshvili, A., Reeves, D., Wellman, M.: Price Prediction Strategies for Market-Based Scheduling. In: 4th Int. Conf. on Automated Planning and Scheduling (ICAPS 2004), pp. 244–252 (2004)
15. Niu, J., Cai, K., Parsons, S., Gerding, E., McBurney, P.: Characterizing Effective Auction Mechanisms: Insights from the 2007 TAC Market Design Competition. In: 7th Int. Conf. on Autonomous Agents and Multiagent Systems (AAMAS 2008), pp. 1079–1086 (2008)
16. Ren, F., Sim, K., Zhang, M.: Market-Driven Agents with Uncertain and Dynamic Outside Options. In: 6th Int. Conf. on Autonomous Agents and Multi-Agent Systems (AAMAS 2007), pp. 721–723 (2007)
17. Sierra, C., Faratin, P., Jennings, N.: A Service-Oriented Negotiation Model between Autonomous Agents. In: Boman, M., Van de Velde, W. (eds.) MAAMAW 1997. LNCS, vol. 1237, pp. 17–35. Springer, Heidelberg (1997)

# The Influence of Culture on ABMP Negotiation Parameters

Gert Jan Hofstede[1], Catholijn M. Jonker[2], and Tim Verwaart[3]

[1] Wageningen University, Hollandseweg 1, 6707 KN Wageningen, The Nethterlands
gertjan.hofstede@wur.nl
[2] Delft University of Tachnology, Mekelweg 4, 2628 CD Delft, The Netherlands
c.m.jonker@tudelft.nl
[3] LEI Wageningen UR, Postbus 29703, 2502 LS Den Haag, The Netherlands
tim.verwaart@wur.nl

**Summary.** Negotiations are known to proceed differently across cultures. A realistic agent model of international negotiations has to take cultural differences into account. This paper presents an agent-based model that tackles this challenge. The context is a trade game where commodities with a hidden quality attribute are exchanged. The negotiation model uses the ABMP negotiation architecture. It applies a utility function that includes market value, quality preference, and risk attitude. The indices of the five dimensions of Hofstede's model of national cultures are used, in combination with agent's group membership and societal status, to differentiate negotiation behavior by adaptation of weight factors in the utility function and ABMP parameters. The paper presents test runs with synthetic cultures and a set of actual national cultures. The present version of the model helps to understand behaviors in international trade networks. It proves that Hofstede's dimensions can be used to generate culturally differentiated agents.

## 1 Introduction

Bargaining practices differ across the world. Multinational companies sometimes work with different price lists for different countries: in order to sell at the same price, the selling company needs to adapt its offer to the varying bargaining practices. A single piece of advice about how to bargain, or a single model to describe bargaining, are obviously not valid across the world unless culture is taken into account.

'Culture' is a notion with many meanings, some of which are contested in some disciplines. However, the leading paradigm today is widely accepted and used in both practice and academia. According to it, culture refers to the *unwritten rules of society*. It is a phenomenon that is specific to a group, not to an individual. It is transmitted in early youth through example and education. As a result its stable across centuries in spite of huge changes in environment and technology. Cultural differences show no signs of diminishing in the Information Age.

T. Ito et al. (Eds.): Innovations in Agent-Based Complex Automated Negotiations, SCI 319, pp. 49–68.
springerlink.com 

Within the literature various basic dimensions can be found according to which societies differ from one another. Of these, the most widely used is Hofstede [1], [2]. His work is accessible, sparse, and based on a very large, very well stratified sample that continues to give it great explanatory value. No other model matches society-level variables so well to date [3].

This paper describes an agent-based model for bargaining in the context of trade and focuses on cultural adaptation of negotiation process parameters. The agents follow common sense strategies such as maximizing utility, seeking good quality, and minimizing risk, but they also have models of how to behave in an appropriate manner. The latter models are based on Hofstede's five dimensions of culture. A cultured agent-based model of bargaining offers several promises. It can help understand the dynamics of international negotiations in trade. It could also serve as a training tool for aspiring international traders.

The paper first briefly introduces Hofstede's model of five dimensions of culture. Next, the ABMP (Agent-Based Market Place, [4]) negotiation model that we adopt is presented. We show how this model can be used in agent-based simulations. We also discuss the limited subset of negotiation situations that are considered in this article. In the third section we link culture and negotiation by describing the influence of each of Hofstede's dimensions of culture on negotiators' practices and preferences. This section sets the scene for the presentation of the rules for our cultured agents in the fourth section. Section five shows example runs with the model on imaginary cultures that have extreme values on one of the Hofstede dimensions and average values on the other four. Section 6 presents results of test runs for a set of actual countries. Finally we discuss the model and how to proceed, since this model forms the basis of future research and tools.

## 2 Hofstede's Five Dimensions of Culture

Human societies have found a different patterns of response to the problems of social life. In some, groups are permanent and close-knit while in others, group membership is volatile and voluntary. In some, leadership style is usually autocratic and in others, participative. Research has shown and repeatedly confirmed that basic tendencies to deal with a few central issues of social life are stable across the generations in societies [2]. They are, because they are instilled into a society's members from birth. As a baby and as a toddler, a child is primed as a social being. Once a child sets foot into the wider society as a teenager, its basic cultural orientation is firmly in place. This research stream has led to dimension models of culture. The most widely used of these is the five-dimension model by Hofstede [1]. The five dimensions are:

- *Identity*: individualism versus collectivism. Essentially this is the extent to which members of a society feel responsible for themselves and close relatives, or for the larger group they belong to, and feel the group to be responsible for them.
- *Hierarchy*: large power distance versus small power distance, or the extent to which the less powerful members of a society expect and accept that power and rights are distributed unequally.

- *Aggression and gender*: masculinity versus femininity. This dimension is about assertive dominance and emotional gender roles. A firm, competitive orientation, versus a consensus-seeking and care-taking orientation, for both women and men.
- *Otherness and truth*: uncertainty avoidance versus uncertainty tolerance. In uncertainty avoiding societies strict rules, rituals, and taboos govern life. Distinctions should be sharp and the unknown is considered dangerous.
- *Gratification of needs*: sort-term versus long-term orientation. In some societies immediate gratification of needs and keeping up social appearance, well-behaving and respect for tradition are seen as virtues. In long-term oriented societies, reasoning is pragmatic and planning, foresight and perseverance are valued.

Note that the five dimensions are not personality traits, but societal patterns. Also note that the picture drawn is necessarily simplified. It presents the two caricatured extremes of each dimension. In reality, almost all cultures have intermediate positions on almost all dimensions. Furthermore, the dimensions of culture can only be isolated from one another in an artificial way. The five dimensions are no more than abstractions that capture main behavioral trends. In reality, cultures have a recognizable feel to them, a Gestalt that can be described, albeit only roughly, by its combination of dimension scores. Experienced negotiators know the range of behaviors that they can expect from negotiators from other parts of the world. They also know how gender, age, status and personality can affect the negotiation style of people from these parts of the world.

In [5], [6], [7], [8], [9], the influence of each of the dimensions on trade processes was modeled separately; a slightly artificial, but also necessary intermediate step to model agents differentiated along the Hofstede dimensions. This chapter describes a reconciled models of all dimensions and presents test runs with synthetic and actual national culture.

## 3 Negotiation Process Model

The present work focuses on a specific type of negotiations: bilateral bargaining about business transactions. The work aims to develop models of actual human behavior. It does not aim to develop an optimal bargaining strategy that can outperform human negotiators or other agents.

In bilateral negotiation, two parties aim at reaching a joint agreement. They do so by exchanging various offers or bids using e.g. an alternating offers protocol [10] called the "negotiation dance" in [11]. Negotiation is a complex emotional decision-making process aiming to reach an agreement to exchange goods or services [12].

The literature on automated negotiation contains a number of agent models for negotiation. The focus of that literature is on reaching deals that are Pareto-efficient (i.e., neither can improve without making the situation worse for the other). Furthermore, some aim at reaching fair outcomes, i.e., in which the deal is equally good for both parties. For more information on strategies see, e.g., [4], [13], [14], [15].

The context of the negotiation behavior modeled in this paper is formed by human gaming simulations. The gaming simulations are designed as tools in supply chains and networks research [16]. The multi-agent model aims to simulate human behavior in these games. Fig. 1 depicts the process model of an agent's activities in the game.

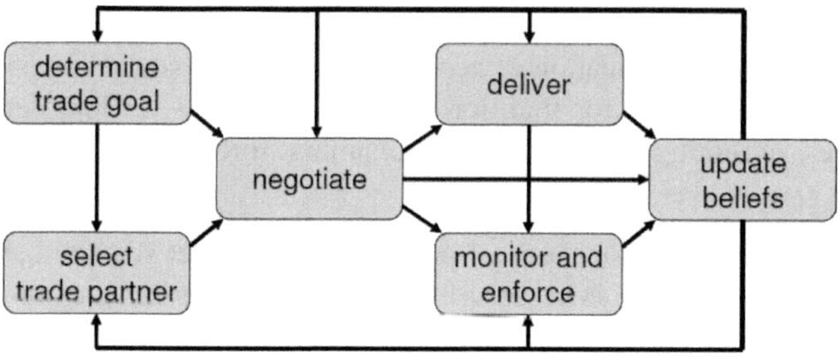


**Fig. 1.** Processes and internal information flows of trading agents

Participants can determine their trade goals, for instance to sell or to buy commodities of a particular quality level. They can select a trade partner according to their preferences and experience, negotiate a transaction of a commodity with quality attributes and guarantees, and deliver truthfully or defect. The essence of the game is information asymmetry: commodities have a quality attribute that is known to the seller and invisible - but testable at some cost - for the buyer. The buyer can either trust the seller's quality statement or spend money on testing, or negotiate some kind of guarantee or certification, which will result in a higher price.

The relevant attributes for comparing bids during the negotiation process are the economic value of the transaction according to market prices, the valuation of particular quality attributes by the trader, and the risk of deceit introduced by the information asymmetry.

The valuations of quality and risk have a rational component and a subjective valuation that is influenced by a trader's personal characteristics and culture. The rational component of a quality attribute is the difference in market price that a trader expects as a result of the quality difference. The rational component of the risk is the product of the amount of the damage and the probability that the damage occurs. The subjective valuation comes in addition to the rational value. For quality, it is the trader's quality preference, for instance because of the societal status that results from trading high quality products. For risk, it is an agent's risk aversion, a cost in excess of *damage* × *probability*, that a trader is willing to make in order to avoid risk.

In the models developed in this work, traders are assumed to compare business proposals by applying a utility function as proposed by Tykhonov et al. [17]:

$$U_B = w_v V_B + w_q Q_B + w_r R_B. \tag{1}$$

$U_B$ stands for the utility that agent $a$ expects from bid $b$ made by partner agent $p$.

$V_B$ reflects $a$'s belief about the economic value of the transaction in the interval [0, 1]. It is calculated as the profit expected from the transaction in case of cooperation, minus the estimated risk of the trade partner to defect, computed as *damage* × *probability*.

$Q_B$ reflects the subjective valuation of the quality attribute of the proposed transaction, in addition to the market value, in the interval [0, 1], e.g. a trader may prefer trading biologically grown food, even if more profit can be made trading traditionally grown.

$R_B$ reflects $a$'s valuation of the risk involved in the interval [0, 1], with 1 representing no risk. It is based on the product of three factors, all normalized values in the interval [0, 1]. The first factor is $a$'s subjective estimate of the probability that $p$'s will defect. The second factor is the opportunity to defect that the contract leaves for $p$, e.g. a contract for organically grown food offers the opportunity to deliver the cheaper traditionally grown, but a contract for traditionally grown food does not. The third factor is the damage that $a$ expects to suffer in case of defection by $p$, normalized in the interval [0, 1] with 1 representing maximal damage. $R_B$ is computed as the product of the three factors.

The factors $w_v$, $w_q$, and $w_r$, with $w_v + w_q + w_r = 1$, reflect the weight that agent $a$ attaches to the terms of the utility function when dealing with $p$. For a rational agent, $w_q = w_r = 0$. The values of $w_q$ and $w_r$ reflect an agent's quality preference and risk aversion, and are to a great extent influenced by culture. Within a culturally homogeneous society, not all agents have equal preferences, but significant differences between cultures exist in the average values of risk aversion and the appraisal of status associated with high quality products.

Technical details of the utility function can be found in [18].

## 4 ABMP Negotiation Architecture

For the agents' negotiation strategy we chose the Agent-Based Market Place (ABMP) architecture of Jonker and Treur [4], because of its proven similarity to human negotiations [19] in experiments with the SAMIN system of which the bidding strategy is that of ABMP.

The ABMP strategy is developed on the following principles. To assess a bid of the other party, it is important to have evaluation methods. Evaluation can be done at two levels: the level of each of the specific attributes (attribute evaluation), and the level of the bid as a whole (overall bid utility). Thus, some characteristics of ABMP are:

- explicit reasoning about the negotiation strategy and co-ordination of the negotiation process;
- evaluation of a bid takes into account both the attributes separately and the overall utility of the bid;
- planning of a new bid takes into account both the overall utility level and the level of attributes separately.

In particular, in the model it is possible to work on two levels: the level of the overall bid and the level of each of the attributes separately. The negotiation model has been specified as a compositional structure within the component Co-operation Management of the agent, see Figure 2.

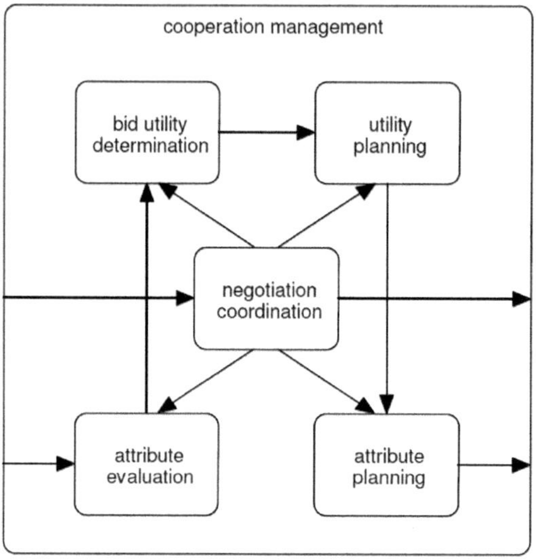


**Fig. 2.** Multi-Attribute Negotiation Architecture

Globally speaking, the process runs as follows:

1. At the beginning of each negotiation round, evaluations of the attributes of the previous bids are determined.
2. The evaluations are aggregated and the overall utilities are computed, both for the own previous bid of the agent and the proposed bid, received from the opponent in the previous negotiation round.
3. Decide whether to accept the opponent's bid, quit the negotiation without a deal, or make a counteroffer. For a counter offer the next step is, step 4.
4. The concession step to be made in the next bid is determined. For this, the agent determines the overall target utility for the next bid to be proposed to the opponent.
5. To obtain the next bid, given the target utility, first according to some distribution over attributes, target attribute evaluation values are determined (chosen in such a manner that they aggregate exactly to the target utility).
6. Finally, for each of these target attribute evaluation values, an attribute value is chosen that has an evaluation value as close as possible to the target evaluation value for the attribute.

An agent accepts another agent's bid, unless:

- *Repetition of steps takes place*: steps without enough progress (depending on the impatience factor ($\pi$) which specifies the acceptable number of steps in which nothing changes)
- A *utility gap* (larger than some threshold $\omega$) remains; i.e., a significant difference between the utility of the own bid and that of the other agent's bid.
- A *configuration mismatch* (larger than some threshold $\nu$) remains between the own bid and the other agent's bid.

Here a configuration mismatch means that for at least one attribute, between the two values (in the two bids) a significant difference exists. Depending on the outcome of the analysis in step 2, component Negotiation Coordination decides on the next action (step 3). If a user is involved, the agent involves the user in the decision process. If the agent is solely responsible for the negotiation, it decides for itself on one of the following actions:

- Start a next negotiation round
- Contact the user (to discuss whether the concession factor ($\gamma$) can be changed.
- Contact the user to discuss whether the configuration tolerance ($\tau$) can be changed.
- Communicate to the user that an agreement has been reached.
- Communicate to the user that the negotiation has failed (only when the user is unwilling to change the characteristics).

One additional elementary aspect of the ABMP architecture is the determination of a concession step (step 4). This step characterizes a concession-based strategy. The decision for what concession to make is based in ABMP on the concept of the current utility gap, i.e., the difference between the utility of the agent's own bid, and the utility of the last bid of the opponent. It is important to note that in this computation both utilities are based on the agent's own utility function. Thus, where the utility of the self (own) bid is the actual, true utility value, the utility of the opponent's bid is typically only an approximation, since the opponent is typically not willing to share his utility function. This illustrates a key aspect of negotiations with incomplete preference information (i.e., where the utility function of the opponent is unknown): the perceived remaining utility gap between the own and other's offer may be very from the perspectives of the two parties. A concession made by one agent (in the way of decreasing his own utility function) may be perceived as no concession at all (even a retraction) by the other party. Similarly, an agent can make no concession (or a very small concession), but the opponent perceives this concession as very considerable. It is precisely this type of concessions that one would like to achieve in multi-attribute negotiations, since it means the new offer is closer to the Pareto-efficient frontier. Such concessions can only be achieved by finding suitable concession trade-offs between the issues.

In ABMP the concession step $CS$ is determined by

$$CS = \beta(1 - \mu/U_{BS})(U_{BO} - U_{BS}). \tag{2}$$

In the above formula, $U_{BS}$ and $U_{BO}$ represent the utilities of self (i.e., the agent's own bid), respectively the utility of the other's last bid. The factor $U_{BO} - U_{BS}$

expresses the current *utility gap*. The letter $\mu$ stands for minimal utility, also known as reservation value. The factor $\beta$ stands for the *negotiation speed*. The minimal utility is taken as $\mu = 1 - \gamma$ with $\gamma$ the *concession factor*, expressing a measure in how far concessions can be made. The factor $(1 - \mu/U_{BS})$ expresses that the concession step will decrease to 0 if the $U_{BS}$ approximates the minimal utility $\mu$. This ensures $U_{BS} \geq \mu$.

Note that in step 4, the user is asked whether he is willing to adjust his concession factor. The rationale for this is that, in human negotiations, often the players do not know their true or absolute cut off value (minimal utility), but start working from a higher value. If it turns out that with that somewhat higher minimal value, they cannot close the deal, humans tend to adjust this minimal value closer to their true cut off value.

The rest of the ABMP architecture is not explained in this paper, as it is not fundamental for understanding the current paper. For more details, the reader is referred to [4] and [20].

# 5 Culture and Bargaining

[5], [6], [7], [8], and [9] model the influence of culture on trade processes for each of the five dimensions separately. Characteristic effects of the dimensions on negotiation behavior are cited below.

## 5.1 Individualism versus Collectivism

According to [5], to a collectivistic trader, negotiation has to be preceded by the formation of a relationship. If that goes wrong there is no negotiation. During the negotiation, collectivistic traders discriminate between in-group and out-group partners. They try to maintain harmony as long as the opponent follows the in-group rules. When doing business with individualist traders the collectivists may be shocked by their opponent's explicit communication. The style of the reaction may be furious, or they might never explicitly say anything, but just avoid the other from now on. In a collectivistic culture the responsibility for in-group welfare and the compliance with in-group rules always play a prominent role. A collectivist will accept benefits for his in-group rather than his personal advantage as a convincing argument.

Individualists have one thing in mind during negotiations: their personal interest. So individualist traders are not very modest in their negotiations, nor will they give in for the purpose of maintaining harmony. If they are not aware of the cultural differences when trading with collectivists, they may be upset by the lack of explicit communication, or they may upset their opponents by being too explicit, or by talking business before the relationship has been established and acknowledged. They behave patiently as long as it serves their interest.

## 5.2 Power Distance

According to [6], traders from egalitarian cultures may have different ways to negotiate, but they will always negotiate. Traders from large power distance cultures on the other hand are not used to negotiating seriously. The powerful dictate the conditions. A trader from a culture with large power distance expects a lower ranked business partner to accept his conditions rapidly. If the lower ranked partner has the same cultural background, there is no problem and the rights of the higher ranked will be recognized and respected: the lower ranked will be modest and give in easily. However, a trader from an egalitarian culture may not give in to the pressure if his status is lower, but will either react furiously (e.g., break off negotiations) or simply ignore the pressure (make a counterproposal), in which case the opponent will be furious (and e.g., break off negotiations).

### 5.3 Masculinity versus Femininity

[7] treat the dimension of masculinity versus femininity as a preference for performance versus cooperation. A performance oriented trader (masculine culture) is interested in fast trades, with as many high quality goods as possible in one trade. This trader is rather impatient, and if bids are too far off from his profile, he will walk away quickly. The performance oriented sticks to the contract of the deal, deceive the trade partner to the limits of the contract without any compunction, and expects the partner to do so too.

A cooperation oriented trader (feminine culture) is interested in the relationship with the trade partner; building trust is important. The amount or quality of goods is not of the most interest, because the relationship built during negotiation might pay off in future negotiations. Given the interest in the relationship with the trade partner, a first negotiation with a trade partner will take time that is willingly spent. During such negotiations, the trader appreciates a negotiation process in which both partners show a willingness to accommodate the other over time.

### 5.4 Uncertainty Avoidance

According to [8], the first bid of an uncertainty avoiding trader tends to be modest in the sense that it is a price he thinks is right. Uncertainty avoiding traders have an emotional style of negotiation, making sure that the opponents understand their feelings. They will not adapt their behavior to their opponent's. In the bargaining that follows they will not easily give in nor will much time be spent. After a few unsuccessful iterations, the uncertainty avoiding trader will break off the negotiation.

Uncertainty tolerant traders on the other hand have a relaxed style of negotiation. They try to adapt their behavior to their counterparts, although they are not prepared to come to an agreement at all cost. They do not show their emotions and may be disconcerted if their opponents do. They are careful not to be more yielding than their counterparts are.

### 5.5 Long Term versus Short Term Orientation

According to [9], long term oriented negotiators are pragmatic and take the bigger picture. They tend to see one bargaining instance as a small step in a long process, and their decisions will be led by their estimation of the profitability or other success chances of that longer process. Long-term oriented traders show patience. They do not break off negotiations. They do not overcharge. A first proposal may be modest, but they do not rapidly give in.

Short term oriented negotiators, on the other hand, think in terms of moral principles and apply them to the situation that is before them here and now. They are very reliable when it comes to following standards of appropriateness of behavior, but this can make them disregard the ulterior consequences of their action.

## 6 Modeling Culture in ABMP

Based on narrative descriptions acquired through an expert systems knowledge elicitation approach, [5], [6], [7], [8], and [9] propose formal models for the influence of the single dimensions on the ABMP parameters, and for the weights that subjective terms for quality preference and risk aversion get in an agent's utility evaluation.

With respect to the influence of culture, the relevant ABMP parameters are concession factor, negotiation speed, utility gap size, and impatience factor. The concession factor determines how far the agent is willing to go in making concessions. Negotiation speed determines the extent of concessions to its own utility the agent would typically make per negotiation round. The utility gap size expresses what is acceptable to the agent when comparing its own bid with that of the opponent. If the difference in utility falls within the utility gap size, the agent will accept the opponent's offer. The impatience factor determines when the agent becomes impatient with the opponent. For example, for some agent it is OK if the other makes a concession within 4 rounds, for another, the opponent should make concessions every round.

[18] proposes a model that integrates the models of the single dimensions into a five-dimensional model of cultural influence. The basis for the integration is the table of cultural factors' influences (Table 1).

The modeling approach is as follows. Behavior is influenced by the cultural dimensions. These are represented by indices usually called PDI, UAI, IND, MAS, and LTO. We scale these indices to the interval [0, 1], 0 representing the lowest observed valuein surveys of national cultures, 1 the highest observed value, and indicate the scaled values by $PDI^*, UAI^*, IND^*, MAS^*$, and $LTO^*$, respectively. So, $IND^*$ indicates the degree of individualism, $1 - IND^*$ indicates the degree of collectivism.

In addition to the indices, in some cultures relational characteristics are relevant to differentiate behavior. In cultures with high power distance, status of the agent and its opponent is relevant. In collectivistic cultures, in-group members are treated differently from out-group members. In uncertainty avoiding cultures, strangers are mistrusted. In short-term oriented cultures, celebrities and well-respected members of society are treated with special attention.

The combination of cultural indices and relational characteristics lead to the cultural factors represented in Table 1. These factors are modeled to have a monotonous effect, either increasing or decreasing, on the relevant ABMP parameters and weight

**Table 1.** Influence of culture on the utility weight factors and ABMP parameters (+ increased parameter value; - decreased; +! increased every negotiation round) [18]

| Culture type | Relation | $w_q$ | $w_r$ | $\gamma$ | $\beta$ | $\omega$ | $\iota$ |
|---|---|---|---|---|---|---|---|
| large power distance | self status high | + | | | | | |
| | self status low | - | | | | | |
| | higher partner | | + | + | | +! | - |
| | lower partner | | - | | | | |
| small power distance | | | | | | | |
| uncertainty avoiding | similar partner | + | + | | + | | + |
| | different partner | + | ++ | | + | | + |
| uncertainty tolerant | | | | | | | |
| individualistic | | | | | | | |
| collectivistic | ingroup partner | | | + | | | - |
| | outgroup partner | | + | | - | | |
| masculine | | + | + | | + | | + |
| feminine | | - | | | - | | - |
| long-term oriented | | - | | | | | - |
| short-term oriented | general | + | | | | | |
| | high partners | + | - | + | | | - |

factors. As no further evidence is available, a simple approach is taken and the effects are assumed to be linear: from some default value to a minimal value if the resulting affect is negative; from the default value to a maximal value if the effect is positive:

$$x = x_T + \frac{\Delta x + |\Delta x|}{2}(x_H - x_T) + \frac{\Delta x - |\Delta x|}{2}(x_T - x_L). \tag{3}$$

In the above equation $x_T$ represent the default value of parameter $x$, $x_H$ the maximal value, $x_L$ the minimal value, and $\Delta x$ the effect of joint cultural factors on $x$.

The effects of the factors on a parameter are computed by subtracting the weak disjunction of the decreasing effects from the weak disjunction of the positive factors. So, positive effects cannot reinforce each other, negative effects cannot reinforce each other, but the strongest negative effect can compensate for the strongest positive effect, vice versa. This can be illustrated by the effect of culture on the impatience parameter:

$$\Delta\iota = \max\left\{UAI^*, MAS^*\right\} - \max\left\{\begin{array}{c} PDI^*(s_p - s_a)(1 - IND^*)(1 - g_{ap}) \\ (1 - MAS^*), LTO^*, (1 - LTO^*)s_p \end{array}\right\}. \tag{4}$$

In this equation $s_a$ and $s_p$ represent the agent's beliefs about own and partner's societal status, respectively, and $g_{ap}$ represent $a$'s belief about group distance with $p$. Effects on other parameters are computed similarly, using Table 1.

## 7 Test Runs with Synthetic Cultures

Table 2 presents results of simulated negotiations, applying Jonker and Treur's ABMP architecture [4]. The negotiations are performed in the simulation environment for of commercial transactions, applied in [6], [7], [8], and [9]. The agents are assigned roles of either suppliers or customers. Agents may select a partner in the opposite role and negotiate about the sale of a commodity that has either high or basic quality. However, quality is not visible without third-party testing, so the buyer of a high quality product has to accept risk, i.e. trust the seller. In the current simulation, agents are neutral with respect to trust, i.e. neither trust nor distrust their trade partners. If they agree on high quality, they implicitly accept the risk of deceit. The percentage of high quality transactions reflects the level of risk that the agents are willing to take. It should be noted that the results are not tuned to realistic situations. The figures should not be taken as absolute values. They show tendencies that emerge from the model.

The results in Table 2 show that in a hierarchical agent society, negotiations succeed more frequently if there is status difference: the higher ranking force the transaction and take risk (high rate of high quality transactions) or force the lower ranking to do so. Egalitarian agents do not accept the risk of deceit.

In uncertainty avoiding agent societies, negotiations fail frequently if the partner is different, i.e. partners do not have common group membership. Negotiations are broken off after a few rounds, because the uncertainty avoiding agents have an urge to proceed ("time is money"). They have a strong preference for high quality commodities. They are willing to take a calculated risk to that end, but only with familiar partners. The uncertainty tolerant agents are more balanced in their judgment of transaction value and risk.

Individualistic agents also do not accept proposals that have too little value or too much risk. Collectivistic agents fail more frequently if they negotiate with out-group partners. With in-group partners, they take their time to negotiate and accept the risk of deceit.

Masculine agents are impatient, break-off frequently, and go for high quality. Feminine agents try to finish the negotiations and take their time for it. Nevertheless, they do not succeed more frequently, because the step size of their concessions is too small.

Long term oriented agents show patience in their negotiations and frequently succeed, but they do not accept risk. Yet they accept high quality transactions, because they take their time to negotiate a price that covers the risk. The sort term oriented are less patient and break off more frequently, but this effect is reduced when they trade with high status partners. They accept risk if they are trading high quality products.

These results comply with the expected behavior of the agents and verify the implementation. However, they do not validate that the implemented model generates believable culturally differentiated agent behavior. For validation of the model, results of extensive simulations with realistic values of cultural parameters should be compared with empirical results from literature. A host of literature on negotiation

**Table 2.** Results of simulated negotiations for extreme settings of culture parameters, i.e. the value for the particular dimension is set to either 0.1 or 0.9, the values for the other dimensions are set to 0.5. Default values of parameters $w_q$, $w_r$, $g$, $w$, $b$, and $i$ are set to 0.2, 0.1, 0.7, 0.2, 0.02, and 0.4, respectively; T: number of successful transactions; P: number of failed negotiations; P: percentage of negotions failed; R: average duration in rounds; Q: percentage of transactions with high quality

| Culture type | Relation | T | F | P | R | Q |
|---|---|---|---|---|---|---|
| large power distance | all status high | 56 | 38 | 40 | 3.6 | 24 |
| | all status low | 60 | 41 | 41 | 3.2 | 0 |
| | higher buyers | 61 | 33 | 35 | 3.3 | 25 |
| | higher sellers | 76 | 39 | 34 | 3.1 | 25 |
| small power distance | | 58 | 56 | 49 | 3.2 | 2 |
| uncertainty avoiding | different partners | 39 | 85 | 69 | 2.6 | 0 |
| | similar partners | 65 | 46 | 41 | 2.9 | 22 |
| uncertainty tolerant | | 48 | 76 | 61 | 2.9 | 1 |
| individualistic | | 56 | 63 | 53 | 3.0 | 0 |
| collectivistic | ingroup partners | 81 | 23 | 22 | 3.4 | 14 |
| | outgroup partners | 35 | 77 | 69 | 3.1 | 0 |
| masculine | | 57 | 55 | 49 | 3.0 | 18 |
| feminine | | 48 | 43 | 47 | 3.7 | 10 |
| long-term oriented | | 71 | 27 | 28 | 3.6 | 16 |
| short-term oriented | general | 40 | 72 | 64 | 3.1 | 13 |
| | high partners | 68 | 51 | 43 | 3.0 | 13 |

in particular countries is available, for instance Adair et al. [21] compare negotiations in France, Russia, Japan, Hong Kong, Brazil, and the United States; Kumar and Worm [22] compare negotiations in China and India.

**Table 3.** Example cultures used in simulations

| culture | *PDI** | *UAI** | *IDV** | *MAS** | *LTO** |
|---|---|---|---|---|---|
| 1 | 0.5 | 0.5 | 0.9 | 0.7 | 0.3 |
| 2 | 0.7 | 0.3 | 0.1 | 0.7 | 0.9 |
| 3 | 0.9 | 0.9 | 0.3 | 0.3 | 0.3 |
| 4 | 0.7 | 0.5 | 0.5 | 0.5 | 0.7 |

The remaining part of this section presents an example of data generated by the model. An agent society of 8 suppliers and 8 customers is given time to trade and negotiate about approximately 100 transactions. All suppliers have equal cultural settings and all customers have equal settings. If agents have equal cultural settings, they are considered in-group. All agents have equal status. Table 3 displays the cultural settings. Culture 1 is modeled after North-American cultures, culture 2 is inspired by China, culture 3 by East-European cultures and culture 4 has similarity with India. Table 4 presents results of the simulations. The results in Table 4

**Table 4.** Example results of a simulation run with default parameter settings as in Table 2 and cultures from Table 3

| variable | supplier culture | customer culture 1 | 2 | 3 | 4 |
|---|---|---|---|---|---|
| Successful transactions | 1 | 61 | 45 | 37 | 69 |
| | 2 | 65 | 90 | 37 | 53 |
| | 3 | 49 | 56 | 59 | 63 |
| | 4 | 58 | 61 | 39 | 69 |
| Percentage failed | 1 | 49 | 57 | 69 | 43 |
| | 2 | 45 | 17 | 70 | 41 |
| | 3 | 61 | 47 | 51 | 41 |
| | 4 | 41 | 41 | 66 | 32 |
| Performance [1]) | 1 | | 0.00 | 0.08 | 0.05 |
| | 2 | 0.06 | | 0.09 | 0.10 |
| | 3 | 0.02 | -0.07 | | 0.02 |
| | 4 | 0.11 | 0.05 | 0.07 | |

1) Performance is computed as average normalized price minus average normalized quality. A high value is an advantage for the suppliers; a low value is advantage for the customers.

demonstrate that in the simulation model, the cultural dimension parameters have their influence. They differentiate aggregate performance in mono-cultural settings as well as in intercultural interactions. However, extensive validation is required on the basis of culture and negotiation literature and experimental data. This paper does not cover such validation. It is subject of the authors' current research.

## 8 Test Runs with Actual National Cultures

Tests have been run with agents configured with actual national cultures. For this purpose, a dataset of cultural dimensions of 62 countries was used (subset of a dataset prepared for a new edition of [2], to appear in 2010). For these cultures, the effect on ABMP parameters can be calculated, for sight validation of the parameter adaptations resulting from Table 1. The effects are presented in the Appendix (Table 6).

For all test runs, the agents were configured homogeneously with the parameter values presented in the Appendix (Table 7). All runs were replicated 10 times. The average results are presented in Table 8 in the Appendix. It should be noted that these results represent modeled behavior in a game, not in actual trade. The results show the number of successful transactions, regardless the profitability of the transactions.

The results show that the model can generate differentiation of behavior in trade for realistic cultural configurations. The effects on different observables, e.g. failure rate and average negotiation duration, are different across cultures. However, more data are required for a thorough analysis, for instance varying the group distance and status differences between agents.

## 9 Discussion and Conclusion

It is well-known that negotiation outcomes differ across the world and that people from different countries differ with respect to the way they negotiate and the results they obtain [23]. Kumar and Worm [22] relate differences in business negotiation processes with differences in economic institutions. According to Hofstede [1], the efficiency of different organizational structures and institutions depend on culture. There is ubiquitous evidence that the result of decision making in business is influenced by the cultural background of the decision makers. As a consequence, realistic business simulation models of international supply chains and networks that take the interaction between business partners into account, should incorporate culture.

Culturally differentiated behavior is not relevant in agent-to-agent negotiations, or other situations where the main purpose of application of intelligent agents is to outperform people by rational decision making, like advocated by Raiffa [11]. Culturally differentiated negotiating agents are useful in a context where human factors play a role. Social simulation is an example of such a context. [17] report a multi-agent simulation that is intended for use in combination with a gaming simulation, as a data gathering tool in supply chain research. Other application area's may be training and education, and decision support systems for human negotiations.

This paper contributes to the understanding of culture's influence on decision making in business by exploring the feasibility of Hofstede's five-dimensional model to simulate believable agents in business. The model has been tested on imaginary cultures that differ on only one of the dimensions.

Data sets have been generated on the basis of actual national cultures. These data show relations between cultural configuration and the simulated frequency of transactions, negotiation failure ratio, duration of negotiation, quality level of traded commodities, and certification level. The results provide some face validity of the simulations, but a deeper analysis on more data is required to reveal relations in the results that can be validated.

Preliminary results of the simulation of more complex, reality-based cultures give evidence that culture in agents can be simulated by applying Hofstede's model, as was originally suggested by de Rosis et al. [24]. However, extensive validations remain for future research. A first source of validation data are the numerous papers reporting differences in negotiations across cultures, e.g. [21]. Gaming simulations like [17] could be used as a tool to collect data for more precise tuning of the model.

## Acknowledgement

This research is supported by the Dutch technology foundation STW, applied science division of NWO and the Technology Program of the Ministry of Economic Affairs. It is part of the Pocket Negotiator project with grant number VICI-project 08075.

## References


1. Hofstede, G.: Culture's Consequences, 2nd edn. Sage Publications, Thousand Oaks (2001)
2. Hofstede, G., Hofstede, G.J.: Cultures and Organizations: Software of the Mind, Third Millennium Edition. McGraw-Hill, New York (2005)
3. Smith, P.: Nations, Cultures, and Individuals: New Perspectives and Old Dilemmas. Journal of Cross-cultural Psychology 35(1), 50–61 (2004)
4. Jonker, C.M., Treur, J.: An agent architecture for multi-attribute negotiation. In: Nebel, B. (ed.) Proceedings Of the 17th International Joint Conference on AI, IJCAI 2001, pp. 1195–1201 (2001)
5. Hofstede, G.J., Jonker, C.M., Verwaart, T.: Individualism and Collectivism in Trade Agents. In: Nguyen, N.T., Borzemski, L., Grzech, A., Ali, M. (eds.) IEA/AIE 2008. LNCS (LNAI), vol. 5027, pp. 492–501. Springer, Heidelberg (2008)
6. Hofstede, G.J., Jonker, C.M., Verwaart, T.: Modeling Power Distance in Trade. In: David, N., Sichman, J.S. (eds.) MABS 2008. LNCS, vol. 5269, pp. 1–16. Springer, Heidelberg (2009)
7. Hofstede, G.J., Jonker, C.M., Meijer, S., Verwaart, T.: Modeling Trade and Trust across Cultures. In: Stølen, K., Winsborough, W.H., Martinelli, F., Massacci, F. (eds.) iTrust 2006. LNCS, vol. 3986, pp. 120–134. Springer, Heidelberg (2006)
8. Hofstede, G.J., Jonker, C.M., Verwaart, T.: Modeling Culture in Trade: Uncertainty Avoidance. In: Proceedings of 2008 Agent-Directed Simulation Symposium (ADS 2008), SCS, San Diego (2008)
9. Hofstede, G.J., Jonker, C.M., Verwaart, T.: Long-term Orientation in Trade. In: Schredelseker, K., Hauser, F. (eds.) Complexity and Artificial Markets. LNEMS, vol. 614, pp. 107–118. Springer, Heidelberg (2008)
10. Osborne, M.J., Rubinstein, A.: A Course in Game Theory. The MIT Press, Cambridge (1994)
11. Raiffa, H.: Negotiation Analysis: The Science and Art of Collaborative Decision Making. Harvard University Press, Cambridge (2002)
12. Thompson, L.L.: The Mind and Heart of the Negotiator, 3rd edn. Pearson Prentice Hall, Upper Saddle River (2005)
13. Faratin, P., Sierra, C., Jennings, N.R.: Negotiation Decision Functions for Autonomous Agents. Int. J. of Robotics and Autonomous Systems 24, 159–182 (1998)
14. Coehoorn, R.M., Jennings, N.R.: Learning an Opponent's Preferences to Make Effective Multi-Issue Negotiation Trade-Offs. In: Proceedings of 6th International Conference on E-Commerce, pp. 59–68 (2004)
15. Hindriks, K., Tykhonov, D.: Opponent Modeling in Automated Multi-Issue Negotiation using Bayesian Learning. In: Proceedings of AAMAS 2008. International Foundation for Autonomous Agents and Multiagent Systems, pp. 331–338 (2008)
16. Meijer, S., Hofstede, G.J., Beers, G., Omta, S.W.F.: Trust and Tracing game: learning about transactions and embeddedness in a trade network. Production Planning and Control 17, 569–583 (2006)
17. Tykhonov, D., Jonker, C., Meijer, S., Verwaart, T.: Agent-Based Simulation of the Trust and Tracing Game for Supply Chains and Networks. J. of Artificial Societies and Social Simulation 11(3) (2008), `http://jasss.soc.surrey.ac.uk/11/3/1.html`
18. Hofstede, G.J., Jonker, C.M., Verwaart, T.: Cultural Differentiation of Negotiating Agents. Group Decis Negot (2010), doi:10.1007/s10726-010-9190-x

19. Bosse, T., Jonker, C.M., van der Meij, L., Treur, J.: Automated Formal Analysis of Human Multi-Issue Negotiation Processes. Multi-Agent and Grid Systems Journal 4, 213–233 (2008)
20. Jonker, C.M., Robu, V., Treur, J.: An Agent Architecture for Multi-Attribute Negotiation Using Incomplete Preference Information. Autonomous Agents and Multi-Agent Systems Journal 15, 221–252 (2007)
21. Adair, W., Brett, J., Lempereur, A., Okumura, T., Shikhirev, P., Tinsley, C., Lytle, A.: Culture and Negotiation Strategy. Negotiation Journal 20, 87–111 (2004)
22. Kumar, R., Worm, V.: Institutional Dynamics and the Negotiation Process: Comparing India and China. International J. of Conflict Management 15, 304–334 (2004)
23. Gelfand, M.J., Brett, J.M.: The Handbook of Negotiation and Culture. Stanford University Press, Stanford (2004)
24. de Rosis, F., Pelachaud, C., Poggi, I.: Transcultural Believability in Embodied Agents: A Matter of Consistent Adaptation. In: Payr, S., Trappl, R. (eds.) Agent Culture, pp. 75–105. Lawrence Erlbaum Associates, Mahwah

# Appendix

**Table 5.** Effects of culture on ABMP parameters based on Table 1 for a sample of 62 countries, computed for group distance 0.5, supplier status 0.3, and customer status 0.7

| national culture (PDI, IDV, MAS, UAI, LTO) | $\Delta w_a$ | $\Delta w_r$ | $\Delta\gamma$ | $\Delta\beta$ | $\Delta\omega$ | $\Delta\iota$ |
|---|---|---|---|---|---|---|
| Argentina (49, 46, 56, 86, 20) | 0.42 | 0.30 | 0.56 | 0.42 | 0.20 | 0.30 |
| Australia (36, 90, 61, 51, 21) | 0.40 | 0.06 | 0.55 | 0.22 | 0.14 | 0.06 |
| Austria (11, 55, 79, 70, 60) | 0.19 | 0.51 | 0.28 | 0.57 | 0.04 | 0.19 |
| Bangladesh (80, 20, 55, 55, 47) | -0.01 | 0.18 | 0.40 | 0.10 | 0.32 | 0.08 |
| Brazil (69, 38, 49, 76, 44) | 0.25 | 0.37 | 0.39 | 0.25 | 0.28 | 0.25 |
| Bulgaria (70, 30, 40, 85, 69) | 0.16 | 0.63 | 0.35 | 0.25 | 0.28 | 0.16 |
| Canada (39, 80, 52, 48, 36) | 0.16 | 0.07 | 0.45 | 0.04 | 0.16 | 0.04 |
| China (80, 20, 66, 30, 87) | -0.21 | 0.57 | 0.40 | 0.26 | 0.32 | -0.21 |
| Chile (63, 23, 28, 86, 31) | 0.14 | 0.38 | 0.48 | 0.14 | 0.25 | 0.14 |
| Colombia (67, 13, 64, 80, 13) | 0.40 | 0.19 | 0.61 | 0.37 | 0.27 | 0.19 |
| Croatia (73, 33, 40, 80, 58) | 0.20 | 0.51 | 0.34 | 0.20 | 0.29 | 0.20 |
| Czech Republic (57, 58, 57, 74, 70) | 0.04 | 0.53 | 0.23 | 0.31 | 0.23 | 0.04 |
| Denmark (18, 74, 16, 23, 35) | -0.19 | -0.12 | 0.46 | -0.61 | 0.07 | -0.61 |
| Estonia (40, 60, 30, 60, 82) | -0.22 | 0.47 | 0.20 | -0.10 | 0.16 | -0.22 |
| Finland (33, 63, 26, 59, 38) | -0.12 | 0.16 | 0.43 | -0.15 | 0.13 | -0.15 |
| France (68, 71, 43, 86, 63) | 0.23 | 0.60 | 0.27 | 0.29 | 0.27 | 0.23 |
| Great Britain (35, 89, 66, 35, 51) | 0.15 | 0.32 | 0.34 | 0.32 | 0.14 | 0.15 |
| Germany (35, 67, 66, 65, 83) | -0.17 | 0.54 | 0.17 | 0.32 | 0.14 | -0.17 |
| Greece (60, 35, 57, 112, 45) | 0.55 | 0.62 | 0.39 | 0.57 | 0.24 | 0.55 |
| Hong Kong (68, 25, 57, 29, 61) | -0.04 | 0.30 | 0.38 | 0.14 | 0.27 | -0.04 |
| Hungary (46, 80, 88, 82, 58) | 0.30 | 0.59 | 0.29 | 0.76 | 0.18 | 0.30 |
| Indonesia (78, 14, 46, 48, 62) | -0.14 | 0.22 | 0.43 | -0.06 | 0.31 | -0.14 |
| India (77, 48, 56, 40, 51) | 0.02 | 0.22 | 0.34 | 0.12 | 0.31 | 0.05 |
| Iran (58, 41, 43, 59, 14) | 0.29 | -0.01 | 0.60 | 0.02 | 0.23 | -0.01 |
| Ireland (28, 70, 68, 35, 24) | 0.44 | 0.15 | 0.53 | 0.36 | 0.11 | 0.15 |
| Israel (13, 54, 47, 81, 38) | 0.28 | 0.38 | 0.43 | 0.28 | 0.05 | 0.28 |
| Italy (50, 76, 70, 75, 61) | 0.14 | 0.48 | 0.27 | 0.45 | 0.20 | 0.14 |
| Japan (54, 46, 95, 92, 88) | 0.07 | 0.87 | 0.27 | 0.68 | 0.22 | 0.07 |
| Korea South (60, 18, 39, 85, 100) | -0.15 | 0.85 | 0.41 | 0.24 | 0.24 | -0.15 |
| Latvia (44, 70, 9, 63, 69) | -0.28 | 0.41 | 0.22 | -0.28 | 0.18 | -0.28 |
| Lithuania (42, 60, 19, 65, 82) | -0.17 | 0.52 | 0.20 | -0.16 | 0.17 | -0.17 |
| Luxembourg (40, 60, 50, 70, 64) | 0.06 | 0.45 | 0.25 | 0.20 | 0.16 | 0.06 |
| Malaysia (104, 26, 50, 36, 41) | -0.11 | 0.09 | 0.41 | 0.00 | 0.40 | 0.00 |
| Mexico (81, 30, 69, 82, 24) | 0.25 | 0.29 | 0.53 | 0.47 | 0.32 | 0.29 |
| Malta (56, 59, 47, 96, 47) | 0.43 | 0.59 | 0.37 | 0.43 | 0.22 | 0.43 |
| Morocco (70, 46, 53, 68, 14) | 0.37 | 0.08 | 0.60 | 0.21 | 0.28 | 0.08 |
| Netherlands (38, 80, 14, 53, 67) | -0.33 | 0.30 | 0.23 | -0.33 | 0.15 | -0.33 |
| Norway (31, 69, 8, 50, 35) | -0.27 | 0.05 | 0.46 | -0.42 | 0.12 | -0.42 |
| New Zealand (22, 79, 58, 49, 33) | 0.25 | 0.11 | 0.47 | 0.16 | 0.09 | 0.11 |
| Pakistan (55, 14, 50, 70, 50) | 0.20 | 0.35 | 0.43 | 0.20 | 0.22 | 0.20 |
| Peru (64, 16, 42, 87, 25) | 0.29 | 0.35 | 0.53 | 0.29 | 0.26 | 0.29 |
| Philippines (94, 32, 64, 44, 27) | 0.07 | 0.13 | 0.51 | 0.28 | 0.38 | 0.13 |
| Poland (68, 60, 64, 93, 38) | 0.45 | 0.50 | 0.43 | 0.57 | 0.27 | 0.50 |
| Portugal (63, 27, 31, 104, 28) | 0.31 | 0.50 | 0.50 | 0.31 | 0.25 | 0.31 |
| Romania (90, 30, 42, 90, 52) | 0.27 | 0.56 | 0.36 | 0.32 | 0.36 | 0.32 |
| Russia (93, 39, 36, 95, 81) | 0.14 | 0.82 | 0.37 | 0.31 | 0.37 | 0.14 |
| South Africa (49, 65, 83, 49, 34) | 0.49 | 0.37 | 0.46 | 0.66 | 0.20 | 0.37 |
| El Salvador (66, 19, 40, 94, 20) | 0.34 | 0.38 | 0.56 | 0.34 | 0.26 | 0.34 |
| Serbia (86, 25, 43, 92, 52) | 0.32 | 0.58 | 0.38 | 0.35 | 0.34 | 0.35 |

**Table 5.** *(continued)*

| | | | | | | |
|---|---|---|---|---|---|---|
| Singapore (74, 20, 48, 8, 72) | -0.24 | 0.28 | 0.40 | -0.04 | 0.30 | -0.24 |
| Slovak Republic (104, 52, 110, 51, 77) | 0.23 | 0.84 | 0.40 | 0.76 | 0.40 | 0.23 |
| Slovenia (71, 27, 19, 88, 49) | 0.07 | 0.52 | 0.37 | 0.07 | 0.28 | 0.07 |
| Spain (57, 51, 42, 86, 48) | 0.28 | 0.50 | 0.36 | 0.28 | 0.23 | 0.28 |
| Sweden (31, 71, 5, 29, 53) | -0.48 | 0.05 | 0.33 | -0.66 | 0.12 | -0.66 |
| Taiwan (58, 17, 45, 69, 93) | -0.24 | 0.64 | 0.42 | 0.14 | 0.23 | -0.24 |
| Thailand (64, 20, 34, 64, 32) | 0.02 | 0.16 | 0.48 | -0.02 | 0.26 | -0.02 |
| Trinidad and Tobago (47, 16, 58, 55, 13) | 0.45 | -0.03 | 0.61 | 0.16 | 0.19 | -0.03 |
| Turkey (66, 37, 45, 85, 46) | 0.30 | 0.47 | 0.38 | 0.30 | 0.26 | 0.30 |
| Uruguay (61, 36, 38, 100, 26) | 0.38 | 0.48 | 0.52 | 0.38 | 0.24 | 0.38 |
| U.S.A. (40, 91, 62, 46, 26) | 0.36 | 0.10 | 0.52 | 0.24 | 0.16 | 0.10 |
| Venezuela (81, 12, 73, 76, 16) | 0.27 | 0.17 | 0.59 | 0.32 | 0.32 | 0.17 |
| Vietnam (70, 20, 40, 30, 57) | -0.17 | 0.10 | 0.40 | -0.20 | 0.28 | -0.20 |

**Table 6.** Default and maximal and minimal values used for cultural adaptations in Table 7

| | $w_q$ | $w_r$ | $\gamma$ | $\beta$ | $\omega$ | $\iota$ |
|---|---|---|---|---|---|---|
| default $x_T$ | 0.05 | 0.05 | 0.6 | 0.3 | 0.02 | 0.3 |
| maximum $x_H$ | 0.2 | 0.2 | 1 | 0.5 | 0.1 | 0 |
| minimum $x_L$ | 0 | 0 | n.a. | 0 | n.a. | 1 |

[the increase of $\omega$ depends on round number $\rho$: $\Delta\omega = PDI^*(s_p\text{-}s_a)\rho$ ];
[the effect of $\iota$=1 is that the probability of quitting a negotiation in case of insufficient progress or unrealistic bidding equals 0.5: $P(\text{quit}) = 0.5\iota$ ]

**Table 7.** Average results of 10 simulation runs with 8 supplier agents and 8 buyer agents, with group distance 0.5, supplier status 0.3 and customer status 0.7; T: number of successful transactions; P: number of failed negotiations; P: percentage of negotions failed; R: average duration in rounds; Q: percentage of transactions with high quality; C: percentage of certified transactions

| national culture (PDI, IDV, MAS, UAI, LTO) | T | F | P | R | Q | C |
|---|---|---|---|---|---|---|
| Argentina (49, 46, 56, 86, 20) | 56 | 36 | 39 | 3.1 | 34 | 7 |
| Australia (36, 90, 61, 51, 21) | 70 | 25 | 26 | 4.3 | 30 | 5 |
| Austria (11, 55, 79, 70, 60) | 60 | 41 | 41 | 4.0 | 25 | 4 |
| Bangladesh (80, 20, 55, 55, 47) | 68 | 38 | 36 | 3.6 | 26 | 5 |
| Brazil (69, 38, 49, 76, 44) | 65 | 26 | 29 | 3.8 | 34 | 5 |
| Bulgaria (70, 30, 40, 85, 69) | 56 | 33 | 37 | 3.3 | 37 | 3 |
| Canada (39, 80, 52, 48, 36) | 62 | 33 | 35 | 3.8 | 34 | 8 |
| China (80, 20, 66, 30, 87) | 87 | 15 | 15 | 4.2 | 20 | 2 |
| Chile (63, 23, 28, 86, 31) | 75 | 18 | 19 | 4.1 | 39 | 7 |
| Colombia (67, 13, 64, 80, 13) | 83 | 30 | 27 | 3.4 | 46 | 7 |
| Croatia (73, 33, 40, 80, 58) | 58 | 37 | 39 | 3.5 | 28 | 5 |
| Czech Republic (57, 58, 57, 74, 70) | 60 | 27 | 31 | 4.2 | 23 | 4 |
| Denmark (18, 74, 16, 23, 35) | 58 | 14 | 19 | 5.3 | 7 | 1 |
| Estonia (40, 60, 30, 60, 82) | 71 | 20 | 22 | 4.7 | 11 | 2 |
| Finland (33, 63, 26, 59, 38) | 69 | 16 | 19 | 4.5 | 23 | 5 |
| France (68, 71, 43, 86, 63) | 63 | 35 | 36 | 3.2 | 21 | 0 |
| Great Britain (35, 89, 66, 35, 51) | 70 | 51 | 42 | 4.0 | 27 | 0 |
| Germany (35, 67, 66, 65, 83) | 73 | 26 | 26 | 4.3 | 19 | 1 |
| Greece (60, 35, 57, 112, 45) | 54 | 38 | 41 | 3.0 | 41 | 9 |
| Hong Kong (68, 25, 57, 29, 61) | 68 | 31 | 31 | 3.8 | 24 | 3 |
| Hungary (46, 80, 88, 82, 58) | 62 | 27 | 30 | 3.6 | 23 | 5 |
| Indonesia (78, 14, 46, 48, 62) | 77 | 25 | 25 | 4.4 | 27 | 6 |

**Table 7.** (*continued*)

| | | | | | | |
|---|---|---|---|---|---|---|
| India (77, 48, 56, 40, 51) | 63 | 26 | 29 | 3.5 | 32 | 9 |
| Iran (58, 41, 43, 59, 14) | 71 | 22 | 24 | 4.0 | 42 | 7 |
| Ireland (28, 70, 68, 35, 24) | 67 | 30 | 31 | 4.4 | 31 | 5 |
| Israel (13, 54, 47, 81, 38) | 52 | 38 | 42 | 3.2 | 27 | 5 |
| Italy (50, 76, 70, 75, 61) | 53 | 34 | 39 | 3.7 | 26 | 6 |
| Japan (54, 46, 95, 92, 88) | 67 | 36 | 35 | 3.8 | 15 | 2 |
| Korea South (60, 18, 39, 85, 100) | 79 | 16 | 17 | 4.2 | 20 | 2 |
| Latvia (44, 70, 9, 63, 69) | 56 | 33 | 37 | 4.2 | 12 | 0 |
| Lithuania (42, 60, 19, 65, 82) | 64 | 31 | 33 | 4.5 | 19 | 1 |
| Luxembourg (40, 60, 50, 70, 64) | 67 | 30 | 31 | 4.0 | 18 | 2 |
| Malaysia (104, 26, 50, 36, 41) | 60 | 41 | 41 | 3.6 | 32 | 2 |
| Mexico (81, 30, 69, 82, 24) | 55 | 45 | 45 | 3.1 | 47 | 11 |
| Malta (56, 59, 47, 96, 47) | 59 | 35 | 37 | 3.2 | 24 | 2 |
| Morocco (70, 46, 53, 68, 14) | 65 | 20 | 24 | 3.6 | 35 | 2 |
| Netherlands (38, 80, 14, 53, 67) | 46 | 31 | 40 | 5.2 | 17 | 0 |
| Norway (31, 69, 8, 50, 35) | 66 | 24 | 27 | 5.3 | 15 | 4 |
| New Zealand (22, 79, 58, 49, 33) | 66 | 40 | 38 | 4.2 | 21 | 4 |
| Pakistan (55, 14, 50, 70, 50) | 72 | 17 | 19 | 4.1 | 29 | 6 |
| Peru (64, 16, 42, 87, 25) | 79 | 24 | 23 | 3.5 | 32 | 2 |
| Philippines (94, 32, 64, 44, 27) | 69 | 31 | 31 | 3.8 | 42 | 12 |
| Poland (68, 60, 64, 93, 38) | 41 | 45 | 52 | 2.5 | 54 | 19 |
| Portugal (63, 27, 31, 104, 28) | 63 | 31 | 33 | 3.4 | 33 | 5 |
| Romania (90, 30, 42, 90, 52) | 68 | 26 | 28 | 3.4 | 37 | 7 |
| Russia (93, 39, 36, 95, 81) | 70 | 27 | 28 | 3.3 | 29 | 6 |
| South Africa (49, 65, 83, 49, 34) | 67 | 37 | 36 | 3.8 | 36 | 7 |
| El Salvador (66, 19, 40, 94, 20) | 69 | 35 | 34 | 3.7 | 36 | 4 |
| Serbia (86, 25, 43, 92, 52) | 56 | 43 | 43 | 3.3 | 52 | 5 |
| Singapore (74, 20, 48, 8, 72) | 75 | 17 | 18 | 4.2 | 20 | 5 |
| Slovak Republic (104, 52, 110, 51, 77) | 75 | 17 | 18 | 3.5 | 24 | 4 |
| Slovenia (71, 27, 19, 88, 49) | 71 | 33 | 32 | 4.2 | 32 | 4 |
| Spain (57, 51, 42, 86, 48) | 61 | 22 | 27 | 4.0 | 26 | 1 |
| Sweden (31, 71, 5, 29, 53) | 51 | 10 | 16 | 7.3 | 4 | 1 |
| Taiwan (58, 17, 45, 69, 93) | 82 | 10 | 11 | 4.2 | 28 | 4 |
| Thailand (64, 20, 34, 64, 32) | 66 | 38 | 37 | 3.5 | 35 | 4 |
| Trinidad and Tobago (47, 16, 58, 55, 13) | 80 | 16 | 17 | 4.0 | 42 | 10 |
| Turkey (66, 37, 45, 85, 46) | 58 | 25 | 30 | 3.7 | 38 | 9 |
| Uruguay (61, 36, 38, 100, 26) | 50 | 33 | 40 | 3.1 | 26 | 9 |
| U.S.A. (40, 91, 62, 46, 26) | 72 | 27 | 27 | 4.4 | 40 | 4 |
| Venezuela (81, 12, 73, 76, 16) | 75 | 24 | 24 | 3.6 | 39 | 5 |
| Vietnam (70, 20, 40, 30, 57) | 69 | 22 | 24 | 4.4 | 22 | 8 |

# Supporting the Design of General Automated Negotiators*

Raz Lin[1], Sarit Kraus[2], Dmytro Tykhonov[3], Koen Hindriks[3], and Catholijn M. Jonker[3]

[1] Department of Computer Science, Bar-Ilan University, Ramat-Gan, Israel 52900
linraz@cs.biu.ac.il
[2] Department of Computer Science, Bar-Ilan University, Ramat-Gan, Israel 52900 and Institute for Advanced Computer Studies, University of Maryland, College Park, MD 20742 USA
sarit@cs.biu.ac.il
[3] Man-Machine Interaction Group, Delft University of Technology, Mekelweg 4, Delft, The Netherlands
{d.tykhonov,k.v.hindriks,C.M.Jonker}@tudelft.nl

**Summary.** The design of automated negotiators has been the focus of abundant research in recent years. However, due to difficulties involved in creating generalized agents that can negotiate in several domains and against human counterparts, many automated negotiators are domain specific and their behavior cannot be generalized for other domains. Some of these difficulties arise from the differences inherent within the domains, the need to understand and learn negotiators' diverse preferences concerning issues of the domain and the different strategies negotiators can undertake. In this paper we present a system that enables alleviation of the difficulties in the design process of general automated negotiators termed GENIUS, a **G**eneral **E**nvironment for **N**egotiation with **I**ntelligent multi-purpose **U**sage **S**imulation. With the constant introduction of new domains, e-commerce and other applications, which require automated negotiations, generic automated negotiators encompass many benefits and advantages over agents that are designed for a specific domain. Based on experiments conducted with automated agents designed by human subjects using GENIUS we provide both quantitative and qualitative results to illustrate its efficacy. Our results show the advantages and underlying benefits of using GENIUS for designing general automated negotiators.

## 1 Introduction

One cannot understate the importance of negotiation and the centrality it has taken in our everyday lives, in general, and in specific situations in particular (e.g., hostage crises [21]). The fact that negotiation covers many aspects of our lives has led to

* This research is based upon work supported in part under NSF grant 0705587 and by the U. S. Army Research Laboratory and the U. S. Army Research Office under grant number W911NF-08-1-0144.

T. Ito et al. (Eds.): Innovations in Agent-Based Complex Automated Negotiations, SCI 319, pp. 69–87.
springerlink.com 

extensive research in the area of automated negotiators, that is, automated agents capable of negotiating with other agents in a specific environment. However, when reviewing many of the agents suggested in the literature (e.g., [3, 5, 19]), one cannot ignore the fact that most of them lack two key fundamental features, which are, to our belief, most important for the design of successful general automated negotiators.

The first problem emerges from the inherent design of the automated negotiator. While humans can negotiate in different settings and domains, when designing an automated agent a decision should be made whether the agent should be a general purpose negotiator, that is, able to successfully negotiate in many settings and domain-independent (e.g., Lin *et al.* [24]), or suitable for only one specific domain (e.g., Ficici and Pfeffer [6] for the Colored Trail domain, or Kraus and Lehmann [20] for the Diplomacy game). There are obvious advantages for an agent's specificity in a given domain. It allows the agent's designer to construct better strategies that could allow it to negotiate better, in comparison to a more general purpose negotiator. However, this is also one of the major weaknesses of these type of agents. With the constant introduction of new domains, e-commerce and other applications, which require negotiations, the generality of an automated negotiator becomes important, as automated agents tailored to specific domain are useless since they cannot be used in the new domains and applications.

The second problem is that automated negotiators should work in open environments. Open environments lack a central mechanism for controlling the agents' behavior, and they may encounter human decision-makers whose behavior is diverse, cannot be captured by a monolithic model, make mistakes, is affected by cognitive, social and cultural factors, etc. [1, 22]. Examples of such environments include online markets, patient care-delivery systems, virtual reality and simulation systems used for training (e.g., the Trading Agent Competition (TAC) [33]).

While the two aforementioned difficulties (and proposed solutions) should be dealt with in more detail, in this paper we do not focus on the design of an efficient automated negotiator; we do not even claim that we have the right "formula" to do so. We do, however, present a tool that aims to help facilitate the *design* and *evaluation* of automated negotiators' strategies. The tool, GENIUS, is a **G**eneral **E**nvironment for **N**egotiation with **I**ntelligent multi-purpose **U**sage **S**imulation. To our knowledge, this is the first tool of its kind that both assists in the *design* of strategies for automated negotiators and also *supports* the evaluation process of the agent. Thus, we believe this tool is very useful for agent designers and can take a central part in the process of designing automated agents. While designing agents can be done in any agent oriented software engineering methodology, GENIUS wraps this in an easy-to-use environment and allows the designers to focus on the development of *strategies* for negotiation in an open environment with multi-attribute utility functions.

GENIUS incorporates several mechanisms that aim to support the design of a general automated negotiator. The first mechanism is an analytical toolbox, which provides a variety of tools to analyze the performance of agents, the outcome of the negotiation and its dynamics. The second mechanism is a repository of domains and utility functions. Lastly, it also comprises repositories of automated negotiators. A comprehensive description of the tool is provided in Section 3.

In addition, GENIUS enables the evaluation of different strategies used by automated agents that were designed using the tool. This is an important contribution as it allows researchers to empirically and *objectively* compare their agents with others in different domains and settings. This is an important contribution with respect to the validation of results reported by researchers with regard to their automated negotiators.

In order to verify its efficacy, GENIUS was introduced to students, who were required to design automated agents for different negotiation tasks. Their agents were evaluated and both quantitative and qualitative results were gathered. A total of 65 automated agents were designed by 65 students. We describe the experimental methodology and results in Section 4. The results support our claim that GENIUS helps and supports the design process of an automated negotiator, from the initial design, through the evaluation of the agent, and re-design and improvements, based on its performance.

We begin by reviewing related research with respect to the design of general automated negotiators.

## 2 Related Work

Research on general agent negotiators has given rise to a broad variety of such agents. The strategies of the agents usually vary from equilibrium strategies, optimal approaches and heuristics. Here we focus in particular on agents that are able to conduct bilateral negotiations with incomplete information. Examples of such general agent negotiators in the literature include, among others, Sycara *et al.* [31], who introduce a generic agent called *Bazaar*, Faratin *et al.* [3], who propose an agent that is able to make trade-offs in negotiations and motivated by maximizing the joint utility of the outcome (that is, the agents are utility maximizers that seek Pareto-optimal agreements), Karp *et al.* [15], who take a game-theoretic view and propose a negotiation strategy based on game-trees, Jonker *et al.* [14], who propose a negotiation model called *ABMP*, and Lin *et al.* [24], who propose an agent negotiator called *QOAgent*. All of these agents are proposed as agent negotiators that perform well in different domains, i.e. are domain-independent; for an example of an agent negotiator targeted at a particular negotiation domain, see Li *et al.* [23]. The motivation for introducing these agents, however, has varied and has related to diverse topics , such as learning in negotiation, the use of various heuristics, or negotiating with humans. Typically, alternating offer protocols are used where agents exchange offers in turn [30], sometimes with minor modifications as for example Lin *et al.* [24] proposed. Lomuscio *et al.* [25] in their work, offer useful classification of types of agent negotiators. Nonetheless, the important issue of the evaluation of agents' strategies and comparing between different strategies even in the same environment has not been adequately addressed by these researchers.

As we argue that it is useful to have a generic environment for designing and evaluating agent negotiators, we briefly review related work that is explicitly aimed at the evaluation of various agent negotiators. Most of the work reported herein concerns the evaluation of various *strategies* for negotiation used by such agents. Although some results were obtained by game-theoretic analysis (e.g. [18, 29]), most results were obtained by means of *simulation* (e.g. [2, 5, 8]). Devaux *et al.* [2] present work comparing agents negotiating in internet agent-based markets. In particular, they compare a strategy of their own agent with behavioral based strategies taken from the literature [3]. The simulations are performed with an abstract domain where agents need to negotiate the price of a product. Similarly, Henderson *et al.* [8] present results of a comparison of various negotiation strategies' performance in a simulated car hire scenario. Finally, Matos *et al.* [27] conducted experiments to determine the most successful strategies using an evolutionary approach in an abstract domain called the *service-oriented domain.*

Even though several of the approaches mentioned use a rather abstract domain with a range of parameters that may be varied, we argue that the focus on a single domain in most simulations is restrictive. A similar argument to this end has been put forward in [12]. The analysis of agent negotiators in multiple domains may significantly improve the performance of such agents. To the best of our knowledge, this is the first time that quantitative and qualitative evidence is presented to substantiate this claim.

Manisterski *et al.* [26] discuss how people who design agent negotiators change their design over time. They study how students changed their design of a trading agent that negotiates in an open environment. After initial design of their agents, human designers obtained additional information about the performance of their agents by receiving logs of negotiations between their agents and agents designed by others. These logs provided the means to analyze the negotiation behavior, and an opportunity to improve the performance of the agents. The GENIUS environment discussed here provides a tool that supports such analysis, subsequent improvement of the design, and structures the enhancement process.

With regard to systems that facilitate the actual design of agents or agent strategies in negotiations, few systems are close to our line of suggested work. Most of the systems that can be somewhat related to the main focus of our paper are negotiation support systems (e.g., the Interactive Computer-Assisted Negotiation Support system (ICANS) [32], the InterNeg Support Program for Intercultural REsearch (INSPIRE)), however, they do not deal with the combination of both the evaluation of strategies and the facilitation of automated negotiator's design. INSPIRE [17] is a Web-based negotiation support system, which primary goal is to facilitate negotiation research in an international environment. The system enables negotiation between two humans and collects data about negotiations and has some basic functionality for the analysis of the agreements, such as calculation of the utility of an agreement and exchanged offers. However, it does not allow integration of an automated negotiating agent and thus does not include repositories of agents as we propose. Perhaps Neg-o-Net [7] is the most similar to GENIUS than all other support systems. The Neg-o-Net model is a generic agent-based computational simulation model for capturing multi-agency negotiations concerning resource and environmental management decisions. Neg-o-Net model includes both negotiation algorithm and agent models. Agent's preferences are modeled using digraphs (scripts). Nodes represent states of the agent that can be achieved by performing actions (arcs). Each state is evaluated using utility functions. The user can modify agent's script to model his/her preferences w.r.t. states and actions. Yet, their system does not allow for the incorporation of human negotiators, but only automated ones. Moreover, they do not provide any evaluation mechanism of the strategies as GENIUS provides.

We continue with a detailed description of the GENIUS system, followed by the experiments we conducted and the results.

## 3 The GENIUS System

GENIUS is a **G**eneral **E**nvironment for **N**egotiation with **I**ntelligent multi-purpose **U**sage **S**imulation. The aim of the tool is to facilitate the design of negotiation strategies. Using GENIUS programmers can focus mainly on the strategy design. This is achieved by GENIUS by providing both a flexible and easy-to-use environment for implementing agents and mechanisms that support the strategy design and analysis of the agents.

GENIUS enables negotiation between automated agents, as well as humans. Human negotiators and automated ones can be joined in a single negotiation session. Human negotiators interact with GENIUS via a graphical user interface (GUI). GUIs included in GENIUS allow the human negotiator to exchange offers with his/her counterpart, to keep track of them, and consult with his/her own preference profile (that is, a utility score assigned to each issue of the negotiation) to evaluate the offers. Figure 1 shows an example of a human negotiator GUI. For automated agents, GENIUS provides skeleton classes to help designers implement their negotiating agents. It provides functionality to access information about the negotiation domain and the preference profile of the agent. An interaction component of GENIUS manages the rules of encounter or protocol that regulates the agent's interaction in the negotiation. This allows the agent designer to focus on the design of the agent, and eliminates the need to implement the communication protocol or the negotiation protocol. Existing agents can be easily integrated in the GENIUS by means of adapters[1].

When designing an automated agent, the designer needs to take into account the environment in which the agent will operate. Then, the agent can be best tailored to the specific environment to achieve the most efficient outcomes. The environment determines several parameters which dictate the number of negotiators taking part in the negotiation, the time frame of the negotiation and the issues on which the negotiation is being conducted. The number of parties participating in the negotiation process can be two (bilateral negotiations) or more (multilateral negotiations). For example, in a market there can be one seller but many buyers, all involved in negotiating over a certain item. On the other hand, if the item is common, there may also be many sellers taking part in the negotiation process.

[1] Indeed as was shown in [10].

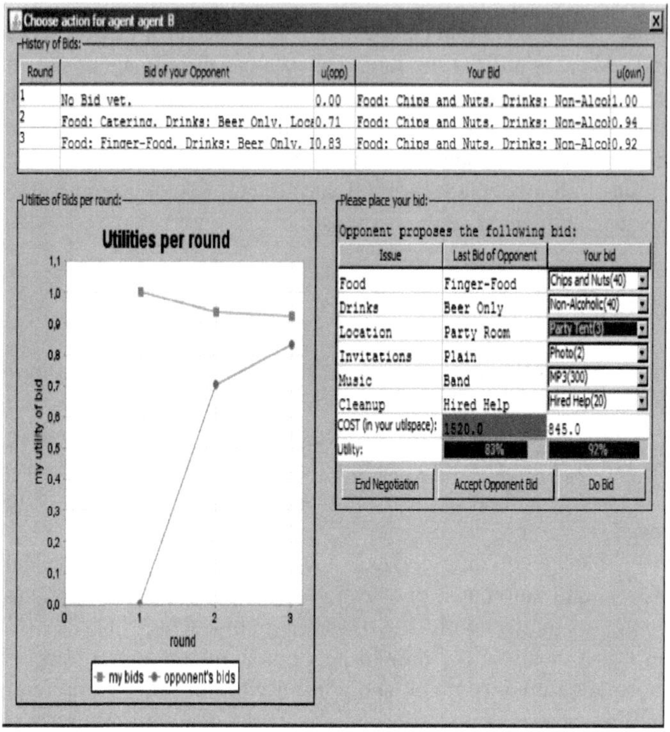


**Fig. 1.** An example of the GUI interface of GENIUS for human negotiators during a specific negotiation session.

The negotiation environment also consists of a set of objectives and issues to be resolved. Various types of issues can be involved, including discrete enumerated value sets, integer-value sets, and real-value sets. A negotiation consists of multi-attribute issues if the parties have to negotiate an agreement which involves several attributes for each issue. Negotiations that involves multi-attribute issues allow making complex decisions while taking into account multiple factors [16]. The negotiation environment can consist of non-cooperative negotiators or cooperative negotiators. Generally speaking, cooperative agents try to maximize their combined joint utilities (e.g., see [34]) while non-cooperative agents try to maximize their own utilities regardless of the other sides' utilities.

Finally, the negotiation protocol defines the formal interaction between the negotiators: whether the negotiation is done only once (one-shot) or repeatedly, and how the exchange of offers between the agents is conducted. A common exchange of offers model is the alternating offers model [30]. In addition, the protocol states whether agreements are enforceable or not, and whether the negotiation has a finite or infinite horizon. The negotiation is said to have a finite horizon if the length of

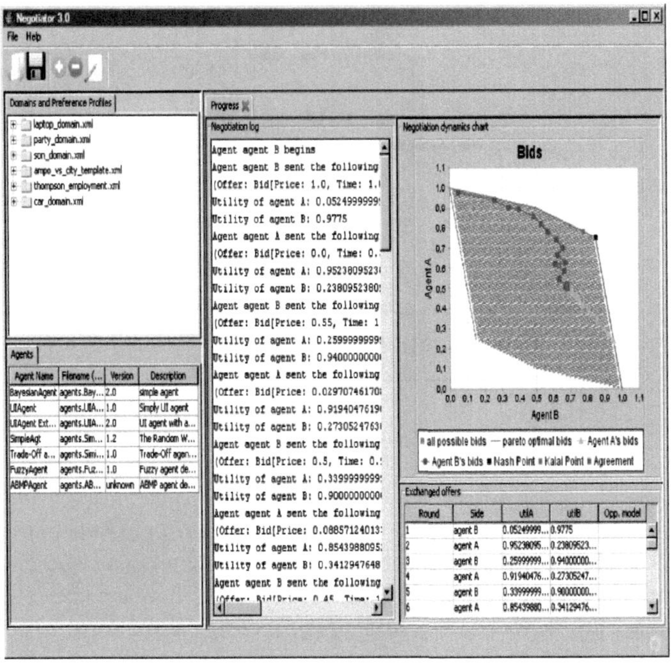


**Fig. 2.** An example of GENIUS' main user interface, showing the results of a specific negotiation session.

every possible history of the negotiation is finite. In this respect, time costs may also be assigned and they may increase or decrease the utility of the negotiator.

Figure 3 depicts the different variations in the settings. GENIUS provide a testbed which allows the designer to easily vary and change these negotiation parameters.

GENIUS provides a flexible simulation environment. A researcher can setup a single negotiation session or a tournament via the GUI simulation (see Figure 2) using the negotiation domains and preference profiles from a repository (top left corner of the GUI simulation), and choose strategies for the negotiating parties (top bottom corner of the GUI simulation). For this purpose, a graphical user interface layer provides options to create a negotiation domain, defines agent preferences, allows human user(s) to participate in a negotiation, and reviews performance and benchmark results of agents that conducted a negotiation. This also includes defining different preferences for each role.

A negotiation domain is a specification of the objectives and issues to be resolved by means of negotiation. Objectives allow to define a tree-like structure with either other objectives again or issues as children. Various types of issues are allowed, including discrete enumerated value sets, integer-valued sets, real-valued sets, as well as a special type of issue called price issue. Additionally, a specification of a negotiation domain may introduce constraints on acceptable outcomes.

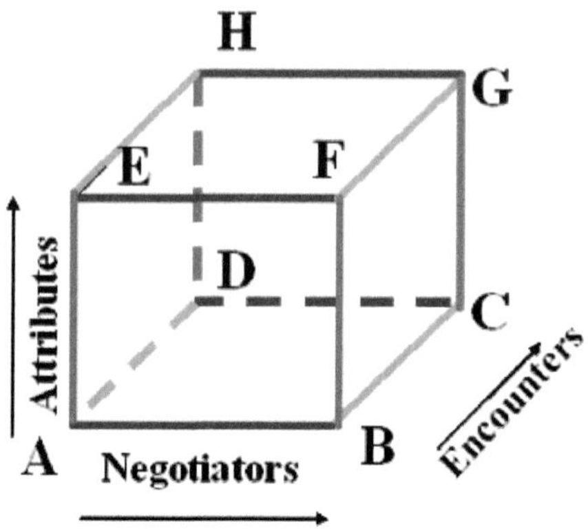


**Fig. 3.** Variations of the negotiation settings.

A preference profile specifies the preferences regarding possible outcomes of an agent. This can be considered a function mapping outcomes of a negotiation domain on the level of satisfaction of an agent associated with that outcome. The structure of a preference profile for obvious reasons resembles that of a domain specification. The tree-like structure enables specification of relative priorities of parts of the tree.

Seven negotiation domains are currently collected in the repository of GENIUS. Each domain has at least two preference profiles required for bilateral negotiations. The number of issues in the domains ranges from 3 to 10, where the largest negotiation domain in the repository is the AMPO vs City taken from [28], and has over 7,000,000 possible agreements. Issues in the repository have different predictabilities of the evaluation of alternatives. Issues are considered predictable when even though the actual evaluation function for the issue is unknown, it is possible to guess some of its global properties (for more details, see [12]). The repository of strategies currently contains six automated negotiation strategies, such as the ABMP strategy [13], the Zero-Intelligence strategy [9], the QO-strategy [24], the Bayesian strategy [11] and others. The repositories of domains and of agents allow agent designers test their agents on the different domains and against different kind of agents and strategies.

GENIUS provides an analytical toolbox for evaluating negotiation strategies. The toolbox calculates optimal solutions, such as the Pareto efficient frontier, Nash product and Kalai-Smorodinsky [28]. These solutions are visually shown to the negotiator or the designer of the automated agent, as depicted in the top right corner of Figure 2. We can see all the possible agreements in the domain (all dotted areas) where the highest and most right lines denote the Pareto efficient frontier. During the negotiation each side can see the distance of its offers from this Pareto frontier as well as the distance from previous offers (as shown by the two lines inside the curve).

Using the analytical toolbox one can analyze the dynamic properties of a negotiation session, such as a classification of negotiation moves (a step-wise analysis of moves) and the sensitivity to a counterpart's preferences measure, as suggested by Hindriks *et al.* [9]. For example, one can see whether his/her strategy is concession oriented, i.e., steps are intended to be concessions, but in fact some of these steps might be *unfortunate*, namely, although from the receiver's perception the proposer of the offer is conceding, the offer is actually worse than the previous offer. The result of the analysis can help agent designers improve their agents.

Moreover, negotiating agents designed using heuristic approaches need extensive evaluation, typically through simulations and empirical analysis, as it is usually hard to predict precisely how the system and the constituent agents will behave in a wide variety of circumstances. To do so, there is a genuine need for the development of a best practice repository for negotiation techniques. That is, a coherent resource that describes which negotiation techniques are best suited to a given type of problem or domain. Repositories of agents and negotiation domains available in GENIUS make it an attractive tool for test bedding negotiating agents. To steer the research in the area of negotiating agents an Automated Negotiating Agents Competition is organized using the GENIUS environment[2].

## 4 Experiments

The experiments described below were conducted in order to test the efficacy of the mechanisms incorporated into GENIUS. Prior to these experiments we verified that GENIUS indeed facilitates the flexible creation of tournaments. As an example, in [9] we evaluated several negotiation strategies in a tournament setup where every negotiation strategy had to negotiate on several different negotiation domains with various preference profiles and against a range of negotiation strategies used by different opponents. As a result, we found that negotiation strategies that are designated as generic and are meant to perform well independent of the domain, nevertheless may be inefficient in particular negotiation setups. For example, the Trade-Off strategy, introduced in [3], shows excellent performance when confronted with itself but its performance is not as good when negotiating against an agent that uses a suboptimal strategy. Furthermore, evidently the characteristics of the negotiation domain and preference profiles, such as the number of issues, the opposition of the preferences and their predictability [9, 12], play a significant role in the performance of negotiation strategies. These results were obtained with the help of the analytical toolbox in GENIUS using GENIUS's repositories of domains, preference profiles, and strategies.

In the experiments we present in this paper, human subjects were instructed to design automated agents that will negotiate with other automated agents in a tournament in an open environment. The experiments were conducted in several phases in order to validate the results. These experiment results show that GENIUS indeed supports the design of general automated negotiators. In the following subsections

[2] For more details on ANAC competition see: http://mmi.tudelft.nl/negotiation/tournament

we describe the negotiation domains, the experimental methodology and we review the results. We begin by presenting the negotiation domains.

## 4.1 Experimental Domain

While the first experiment was only run on one domain, the second experiment was run on three domains. In the first two domains we modeled three possible agent types, and thus a set of six different utility functions was created for each domain. In the third domain only one type was possible for the different roles. The different types of agents describe the different approaches towards the negotiation process and the other party. For example, the different approaches can describe the importance each agent associates with the effects of the agreement over time. One agent might have a long term orientation regarding the final agreement. This type of agent would favor agreements concerned more with future outcomes of the negotiations, than those focusing only on solving the present problem. On the other hand, another agent might have a short term orientation which focuses on solving only the burning issues under negotiation without dealing with future aspects that might arise from the negotiation or its solutions. Finally, there can also be agents with a compromise orientation. These agents try to find the middle grounds between the possible agreements.

Each negotiator was assigned a utility function at the beginning of the negotiations but had incomplete information regarding the counterpart's utility. That is, the different possible types of the counterpart were public knowledge, but the exact type of the counterpart was unknown.

We describe the three domains in the following subsections. The first two domains are taken from [24], in which they were used for negotiations by human negotiators as well as automated ones. The third domain is taken from the Dispute Resolution Research Center at Kellogg School of Management.

### 4.1.1 The World Health Organization's Framework Convention on Tobacco Control Domain

In this scenario England and Zimbabwe negotiate in order to reach an agreement evolving from the World Health Organization's Framework Convention on Tobacco Control, the world's first public health treaty. The principal goal of the convention is "to protect present and future generations from the devastating health, social, environmental and economic consequences of tobacco consumption and exposure to tobacco smoke."

The leaders of both countries are about to meet at a long scheduled summit. They must reach an agreement on 4 issues, each with several attributes:

1. The total amount to be deposited into the Global, Tobacco Fund to aid countries seeking to rid themselves of economic dependence on tobacco production;
2. Impact on other aid programs;
3. Trade issues;

4. Creation of a forum to explore comparable arrangements for other long-term health issues.

Consequently, a total of 576 possible agreements exist in this domain. While for the first two issues there are contradictory preferences for England and Zimbabwe, for the last two issues there are options which might be jointly preferred by both sides.

### 4.1.2 The Job Candidate Domain

In this scenario, a negotiation takes place after a successful job interview between an employer and a job candidate. In the negotiation both the employer and the job candidate wish to formalize the hiring terms and conditions of the applicant. In contrast to the England-Zimbabwe scenario, some issues *must* be agreed upon to achieve even a partial agreement. Below are the issues under negotiation:

1. Salary;
2. Job description;
3. Social benefits;
4. Promotion possibilities;
5. Working hours.

In this scenario, a total of 1,296 possible agreements exist.

### 4.1.3 The Class Project Domain

In this scenario, Bob and Alice need to decide on a final project plan. In contrast to the other two domains, in this domain the utility preferences of both sides are completely symmetric. For each issue, five possible values are negotiable. The issues under negotiation are:

1. Project's topic;
2. Project's type;
3. Method of presentation;
4. Completion time;
5. Preparation time;
6. Meeting times.

This is also the largest scenario of all three, in terms of possible agreements. In this scenario, a total of 15,625 possible agreements exist. Yet, unlike the previous domains, only one type for each role was possible.

## 4.2 Experimental Methodology

We evaluated the process of the agents design by requiring computer science undergraduate and graduate students to design automated agents. These agents were matched twice in a tournament with all other agents. After each tournament, the students were exposed to one of the mechanisms of GENIUS and were allowed to re-design their agent. Then, they were matched again in a tournament. In addition, after the students submitted their new agents, they were required to fill in questionnaires and evaluate the design process of their agents.

We conducted two experiments. In the first, we evaluated the efficacy of the analytical toolbox. The second experiment was designed to enable evaluation of the efficacy of the repositories of domains and repositories of agents. We describe both experiments in the following subsections.

### 4.2.1 Evaluation of the Analytical Toolbox

In the first experiment, 51 undergraduate students were required to design an automated negotiator using the GENIUS environment. The students were instructed to design an automated negotiator which will be able to negotiate in several domains, however, they were only given the Job Candidate domain described in Section 4.1.2 as an example. In addition, three automated negotiators were supplied with the tool[3]:

1. An agent that follows the Bayesian strategy [11];
2. Another automated agent that follows the Agent-Based Market Places (ABMP) strategy, which is a concession-oriented negotiation strategy [13], though, the strategy itself was not explained to the students;
3. A simple agent that sorts all possible offers according to their utility and sends them one-by-one to the opponent starting with the highest utility.

In the first phase, the students were unaware of the analytical toolbox (which was also removed from the environment and the code). After the students submitted their agent, they were given an upgraded environment which included the analytical toolbox. They were given an explanation about its features. Then they were allocated several days in which they could use it to re-design their agent.

The students' agents were evaluated three times. The first time included running the first phase agents against all other agents. Thus, each agent was matched against all 51 agents (including itself), each time under a different role. That is, each agent participated in 102 negotiations, and a total of 5,202 simulations were executed. The second time, each revised agent was matched against all 51 revised agents (including itself). This allowed us to validate the efficacy of the analytical toolbox by comparing the performance of each revised agent to its original performance. The third time included running the revised agents against each other using a new domain, the England-Zimbabwe domain, which they were unaware of during the design process. This allowed us to evaluate whether the analytical toolbox by itself is or is not suffice for designing generalized agents.

### 4.2.2 Evaluation of the Domain and Agent Repositories

In this experiment, like the previous experiment, 14 graduate students were required to design an automated negotiator using the GENIUS environment. They were also

[3] The agents were supplied with their code to also demonstrate to the students the use of skeleton classes.

instructed that their task is to design an efficient negotiator that will be matched with all other automated negotiators. Throughout the design process they were unaware of the analytical toolbox. In the first part of the exercise they were given the Job Candidate domain as an example. After their submissions, they were given an additional domain, the England-Zimbabwe domain described in Section 4.1.1. As in the previous experiment, they were allocated several days in which they could re-design their agents based on the new introduced domain. Furthermore, half of the students were given logs of all their matches during the tournament. The logs included detailed information of the negotiation process.

In this experiment the students' agents were evaluated four times. The first time included running the first phase agents against all other agents. Thus, each agent was matched against all 14 agents (including itself). The agents were run twice. Once on the domain that was known to them during the design of the original agents, i.e., the Job Candidate domain, and once in the England-Zimbabwe domain which they were unaware of at the time. The second time, each revised agent was matched against all 14 revised agents in the Job Candidate domain and in the England-Zimbabwe domain, respectively. This allowed us to validate the efficacy of both the introduction of a new domain and the usage of logs of past negotiations by comparing the performance of each revised agent to its original performance. Lastly, we ran the students' agents against each other using a new domain, the Class Project domain, which the designers were unaware of during the entire design process. Again, we ran both the original agents and the revised agents. This allowed us to evaluate whether or not the two given domains were suffice for designing efficient generalized agents.

### 4.3 Experimental Results

The main goal of the experiments was to verify that the mechanisms in GENIUS assist in alleviating the difficulties in designing efficient general automated negotiators.

As we mentioned earlier, we experimented in three distinct domains. The utility values ranged from -575 to 895 for the England role and from -680 to 830 for the Zimbabwe role; in the Job Candidate domain from 170 to 620 for the employer role and from 60 to 635 for the job candidate role, and in the Class Project domain from 0 to 29,200 for both sides.

#### 4.3.1 Experiments with the Analytical Toolbox

We evaluate the design of the agents using both quantitative results and qualitative results. The quantitative results, presented in Table 1, comprise a comparison of the agents' performance in the different settings of the experiments, while the qualitative results are gathered from the questionnaires the subjects filled in after the submission of the revised agents.

The average utility gained by all the revised agents was 525 when playing the role of the employer and 505 when playing the role of the job candidate. These

**Table 1.** Average utility values gained by the automated agents before and after being exposed to the analytical toolbox.

| Approach/Role | Employer | Job Candidate |
|---|---|---|
| Original Agents | 517 | 490 |
| Revised Agents | 525 | 505 |

averages are significantly higher (using t-test with $p$-value $< 0.001$) in both roles as compared to the average utilities of the original agents (517 and 490, respectively).

In order to assess the ease of use of the GENIUS environment in creating generalized agents, as well as how helpful the analytical toolbox was, the students were asked to answer several questions on a questionnaire they were administered. 67% of the students indicated that they re-designed their agent in the second part, after being introduced to the analytical toolbox, and 79.6% used it to gain a better understanding of the negotiation and to redesign their agents. Moreover, in a scale of 1 (being the lowest) to 7 (being the highest), the students rated the helpfulness of the tool in understanding the dynamics of the negotiation and the strategy of their agent with an average of 4.06. The students indicated that the tool enabled them to attain a clearer view of the negotiation dynamics by visualizing the spectrum of offers and their utilities, and understand which offers to accept and which offers to propose. Some students also commented that the tool helped them verify that their implemented strategy was indeed as they had intended it to be. Figure 4 presents the total rating the students gave for the helpfulness of the analytical toolbox.

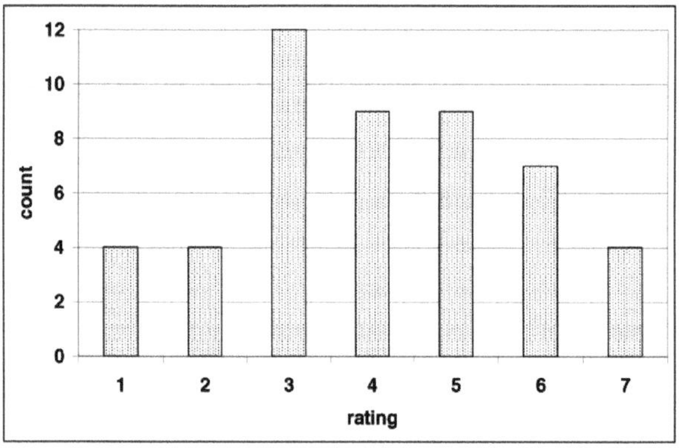


**Fig. 4.** Rating of the helpfulness of the analytical toolbox.

It is interesting to note that most students indicated that they designed their agent to play as if they were the negotiator (an average score of 4.54), yet they also indicated that the fact that they knew that their counterpart would be a computer agent and not a human affected their strategy as they tried to take advantage of this fact.

While this encouraged us as to the efficacy of the analytical toolbox as a supporting mechanism for designing automated negotiators, we still had to verify whether it could also assist in the design of generalized automated negotiators. To test the generality of the agents, we ran the revised agents in a new domain, the England-Zimbabwe domain, of which the students were unaware. However, in this domain only 32.3% of the negotiations were completed successfully, i.e., with a full agreement, as compared to almost double the amount of negotiations that were completed successfully on the known domain (64.4%). That is, while the analytical toolbox was indeed helpful to the students and assisted them in the design of their agent, it was not suffice in order to help them design an efficient general agent. Thus, we continued to devise a second experiment with repositories of domains and agents. The results of this experiment are described in the next subsection.

### 4.3.2 Experiments with Repositories of Domains and Agents

We continued to test other aspects of GENIUS to see whether they help in the design process of agents' strategies. In this experiment, the domains also had a time effect. That is, costs were assigned to each agent, such that during the negotiation process, the agents might gain or lose utility over time. The results are summarized in Tables 2 and 3. In the first part, the students were required to design a general agent, however only one domain was given to them. The average utility scores of their agents in the Job Candidate domain were 363 for the Employer role and 336.8 for the Job Candidate role. In order to evaluate the improvement of the agents due to the logs of past negotiations in which they were matched with all other agents, we continued to run the students' revised agents in the same domain. The results of the agents in this experiment were better, yet not statistically significant (an average utility of 384.29 with a $p$-value $< 0.07$ and 365.78 with a $p$-value $< 0.06$ for the Employer and the Job Candidate roles, respectively). In addition, significantly more negotiations ended with a full agreement (77.3% in the first stage, as compared to 85% in the second stage, $p$-value $< 0.05$).

With respect to using the repositories of agents as a means of improving an agent's strategy, 80% of the students who received the logs of their agents' past negotiations indicated that they indeed used it to improve their agents' behavior. Some noticed, thanks to the logs, that they had bugs in their strategy or that their agents' behavior was too strict and less compromising, causing too many negotiations to end with opting-out. Using this insight, they revised their agents' behavior.

To evaluate the benefits of the repositories of domains on the performance of their agent, we first matched the students' original agents against each other in the new England-Zimbabwe domain. Recall that the original agents were designed without knowledge about the new domain. We then compared these results with the results of the revised agents that had knowledge of the new domain. The average utility scores of the original agents were 302.11 for the England role and -413.57 for the

**Table 2.** Average utility values gained by the automated agents before and after being exposed to logs of past negotiations.

| Approach/Role | Employer | Job Candidate |
|---|---|---|
| Original Agents | 363 | 336.8 |
| Revised Agents | 384.29 | 365.78 |

**Table 3.** Average utility values gained by the automated agents before and after being exposed to an additional domain.

| **England-Zimbabwe Domain** | | |
|---|---|---|
| Approach/Role | England | Zimbabwe |
| Original Agents | 302.11 | -413.57 |
| Revised Agents | 369.99 | -377.37 |
| **Class Project Domain** | | |
| Approach/Role | Bob | Alice |
| Original Agents | 11,357 | 10,655 |
| Revised Agents | 13,348 | 12,113 |

Zimbabwe role. The results of the revised agents were significantly better in the case of England (an average utility of 369.99 with a $p$-value $< 0.03$), while the utility was better, though not statistically significant, for the role of Zimbabwe (-377.37). However, with the revised agents significantly more negotiations ended with a full agreement (39.2% in the first stage, as compared to 50.5% in the second stage, $p$-value $< 0.02$).

To validate these results, the students' agents were then run in the Class Project domain, described in Section 4.1.3, of which they were unaware during their entire design process. We first ran the original agents in that domain, and the average utility scores of the agents were 11,357 for Bob's role and 10,655 for the Alice's role. In addition, only 66.5% of the negotiations ended with a full agreement. We then ran their revised agents against themselves. Consequently, significantly more negotiations ended with a full agreement (76.8%, $p$-value $< 0.02$), resulting also in higher average utility values of 13,348 for Bob and 12,113 for Alice. When the agents played the role of Bob these results were also significant ($p$-value $< 0.04$). We believe that if we had more students' designed agents the average utility values the agents achieved could have been significantly better in both roles, both in the Class Project domain and in the England-Zimbabwe domain.

In this set of experiments we also gave the students questionnaires to help qualitatively assess the efficiency of the repositories of domains and agents. The students had to rate several statements in a scale of 1 (being the lowest) to 7 (being the highest). The students indicated that their agent was more generic after the second domain was introduced. The average score for the agent's generality in the first stage was 5.38 compared to 6.08 for the revised version. Overall, the students rated their agents' generality as 6.0, and they asserted that their agents would succeed in playing well in other domains as well, with an average rating of 5.38.

## 5 Conclusions

Availability of efficient general automated negotiators has two main advantages. Firstly, it minimizes the effort required for adaptation of a general automated negotiator to a new domain. Furthermore, the general automated negotiator can be used as a starting point to create a more efficient negotiator that takes into account a domain specific knowledge, e.g., available a priori information about the most likely preferences of the opponent. Secondly, a general automated negotiated agent is not biased towards domain specific features that can have negative influence on its negotiation efficiency.

This paper presents a simulation environment which supports the design of general automated negotiators. Extensive simulations with more than 60 computer science students were conducted to validate the efficacy of the simulation environment. The results show that GENIUS indeed supports the design of general automated negotiators, and even enables the designers to improve their agents' performance while retaining their generality. This is an important feature, since most of the time general automated negotiators are perceived to perform worse than agents designed specifically for a given domain.

We conducted experiments with automated agents in three distinct domains. The largest domain comprised more than 15,000 possible agreements. While this proves that the simulation environment supports repositories of domains, we did not evaluate the agents on very large domains (e.g., more than 1,000,000 agreements). Many of the automated agents the students designed took advantage of the small domains and reviewed all possible agreements. This would be infeasible in larger domains with a deadline for the negotiation or each turn in the negotiation.

Another issue for future research is the use of GENIUS for the design of automated negotiators that can successfully and efficiently negotiate with human negotiators. As we mentioned, some of the students took advantage of the fact that they were aware that their agents would be matched only with other automated agents. It would be interesting to evaluate the performance of their agents against human negotiators as well.

In future work, we plan to run complete tournaments between the agents in the repository on all available negotiation domains. This would allow us to identify the most efficient strategy currently available in the repository. In addition, we believe that efficiency of a negotiation strategy can depend on the opponent's strategy as well as on the characteristics of the negotiation domain and preference profiles. The analytical toolbox of GENIUS would allow us to identify such dependencies and understand the reasoning behind them. Logs of negotiation sessions produced by GENIUS can be used to discover patterns of negotiation behavior of the automated negotiation strategies of human negotiators.

We also plan to use GENIUS as a training environment to teach people negotiation concepts, such as exploration of outcome spaces, analysis of opponent's offers, making of trade-offs between issues, using concession tactics, etc. Another research direction includes the extension of GENIUS to allow for argumentation and explanation, by allowing the agents to explain its actions to people.

## References

1. Bazerman, M.H., Neale, M.A.: Negotiator rationality and negotiator cognition: The interactive roles of prescriptive and descriptive research. In: Young, H.P. (ed.) Negotiation Analysis, pp. 109–130. The University of Michigan Press (1992)
2. Devaux, L., Paraschiv, C.: Bargaining on an internet agent-based market: Behavioral vs. optimizing agents. Electronic Commerce Research 1, 371–401 (2001)
3. Faratin, P., Sierra, C., Jennings, N.R.: Negotiation decision functions for autonomous agents. Int. Journal of Robotics and Autonomous Systems 24(3-4), 159–182 (1998)
4. Faratin, P., Sierra, C., Jennings, N.R.: Using similarity criteria to make trade-offs in automated negotiations. AIJ 142(2), 205–237 (2002)
5. Fatima, S.S., Wooldridge, M., Jennings, N.R.: A comparative study of game theoretic and evolutionary models of bargaining for software agents. Artificial Intelligence Review 23(2), 187–205 (2005)
6. Ficici, S., Pfeffer, A.: Modeling how humans reason about others with partial information. In: Proceedings of AAMAS 2008, pp. 315–322 (2008)
7. Hales, D.: Neg-o-net - a negotiation simulation test-bed. Technical Report CPM-02-109, CPM (April 2002)
8. Henderson, P., Crouch, S., Walters, R.J., Ni, Q.: Comparison of some negotiation algorithms using a tournament-based approach. In: Agent Technologies, Infrastructures, Tools, and Applications for E-Services, pp. 137–150 (2008)
9. Hindriks, K., Jonker, C., Tykhonov, D.: Negotiation dynamics: Analysis, concession tactics, and outcomes. In: Proceedings of IAT 2007, pp. 427–433 (2007)
10. Hindriks, K., Jonker, C., Tykhonov, D.: Towards an open negotiation architecture for heterogeneous agents. In: Klusch, M., Pěchouček, M., Polleres, A. (eds.) CIA 2008. LNCS (LNAI), vol. 5180, pp. 264–279. Springer, Heidelberg (2008)
11. Hindriks, K., Tykhonov, D.: Opponent modelling in automated multi-issue negotiation using bayesian learning. In: Proceedings of AAMAS 2008, pp. 331–338 (2008)
12. Hindriks, K., Tykhonov, D.: Towards a quality assessment method for learning preference profiles in negotiation. In: Proceeding of Workshop on Agent-based Complex Automated Negotiations, pp. 64–71 (2008)
13. Jonker, C., Treur, J.: An agent architecture for multi-attribute negotiation. In: Proceedings of IJCAI 2001, pp. 1195–1201 (2001)
14. Jonker, C.M., Robu, V., Treur, J.: An agent architecture for multi-attribute negotiation using incomplete preference information. JAAMAS 15(2), 221–252 (2007)
15. Karp, A.H., Wu, R., Yut Chen, K., Zhang, A.: A game tree strategy for automated negotiation. Technical Report HPL-2003-154 (2003)
16. Keeney, R., Raiffa, H.: Decisions with Multiple Objective: Preferences and Value Trade-offs. John Wiley & Sons, NY (1976)
17. Kersten, G.E., Noronha, S.J.: Www-based negotiation support: design, implementation, and use. Decision Support Systems 25(2), 135–154 (1999)
18. Kraus, S.: Strategic Negotiation in Multi-Agent Environments. The MIT Press, Cambridge (2001)

19. Kraus, S., Hoz-Weiss, P., Wilkenfeld, J., Andersen, D.R., Pate, A.: Resolving crises through automated bilateral negotiations. AIJ 172(1), 1–18 (2008)
20. Kraus, S., Lehmann, D.: Designing and building a negotiating automated agent. Computational Intelligence 11(1), 132–171 (1995)
21. Kraus, S., Wilkenfeld, J., Harris, M.A., Blake, E.: The hostage crisis simulation. Simulation & Gaming 23(4), 398–416 (1992)
22. Lax, D.A., Sebenius, J.K.: Thinking coalitionally: party arithmetic, process opportunism, and strategic sequencing. In: Young, H.P. (ed.) Negotiation Analysis, pp. 153–193. The University of Michigan Press (1992)
23. Li, C., Giampapa, J., Sycara, K.: A review of research literature on bilateral negotiations. Technical Report CMU-RI-TR-03-41 (2003)
24. Lin, R., Kraus, S., Wilkenfeld, J., Barry, J.: Negotiating with bounded rational agents in environments with incomplete information using an automated agent. AIJ 172(6-7), 823–851 (2008)
25. Lomuscio, A., Wooldridge, M., Jennings, N.R.: A classification scheme for negotiation in electronic commerce. In: Proceedings of AMEC 2001, pp. 19–33 (2001)
26. Manistersky, E., Lin, R., Kraus, S.: Understanding how people design trading agents over time. In: Proceedings of AAMAS 2008, pp. 1593–1596 (2008)
27. Matos, N., Sierra, C., Jennings, N.R.: Determining successful negotiation strategies: An evolutionary approach. In: Proceedings of ICMAS 1998 (1998)
28. Raiffa, H.: The Art and Science of Negotiation. Harvard University Press, Cambridge (1982)
29. Rosenschein, J., Zlotkin, G.: Designing Conventions for Automated Negotiation among Computers. The MIT Press, Cambridge (1994)
30. Rubinstein, A.: Perfect equilibrium in a bargaining model. Econometrica 50(1), 97–109 (1982)
31. Sycara, K., Zeng, D.: Bayesian learning in negotiation. In: Working Notes of the AAAI Spring Symposium on Adaptation, Co-evolution and Learning in Multiagent Systems (1996)
32. Thiessen, E.M., Loucks, D.P., Stedinger, J.R.: Computer-assisted negotiations of water resources conflicts. GDN 7(2), 109–129 (1998)
33. Wellman, M.P., Greenwald, A., Stone, P.: Autonomous Bidding Agents: Strategies and Lessons from the Trading Agent Competition. MIT Press, Cambridge (2007)
34. Zhang, X., Lesser, V., Podorozhny, R.: Multi-dimensional, multistep negotiation for task allocation in a cooperative system. JAAMAS 10(1), 5–40 (2005)

# Common Testbed Generating Tool Based on XML for Multiple Interdependent Issues Negotiation Problems

Katsuhide Fujita[1], Takayuki Ito[2], and Mark Klein[3]

[1] School of Techno-Business Administration Nagoya Institute of Technology, Gokiso, Showa-ku, Nagoya 466-8555, Japan
fujita@itolab.mta.nitech.ac.jp

[2] School of Techno-Business Administration/Dept. of Computer Science, Nagoya Institute of Technology Gokiso, Showa-ku, Nagoya 466-8555, Japan / Sloan School of Management, Massachusetts Institute of Technology, Three Cambridge Center, NE20-336 Cambridge, MA 02142, USA
ito.takayuki@nitech.ac.jp

[3] Sloan School of Management, Massachusetts Institute of Technology, Three Cambridge Center, NE20-336 Cambridge, MA 02142, USA
m_klein@mit.edu

**Summary.** Multiple interdependent issues negotiations have been widely studied since most real-world negotiation involves multiple interdependent issues. Our work focuses on negotiation with multiple interdependent issues in which agent utility functions are nonlinear. In the field of multiple issue negotiations, there are no established common testbeds for evaluating protocols. In this paper, we propose a common testbed creating tool based on XML that mainly covers the utility functions based on cube-constraints and cone-constraints. First, we propose a testbed generating tool that inputs configuration data and outputs XML formatted files that represent agent utility spaces. The current tool can produce four types of utility spaces: Random, A Single Hill, Two-Hills, and Several Hills. These types are observed in real negotiation settings. Also we define the agent's utility space information based on XML tags. By defining the testbed data as XMLs, users can easily read the files and change the data structure. Finally, we demonstrate experimental results when the existing protocols employ our proposed testbed. Moreover, we introduce some example search programs using our testbeds to evaluate their effectiveness.

## 1 Introduction

Multi-issue negotiation protocols represent an important field of study. While there has been a lot of previous work in this area [2, 3], most of it deals exclusively with simple negotiations involving independent multiple issues. Many real-world negotiation, however, are complex ones involving interdependent multiple issues. Thus, we focus on complex negotiation with interdependent multiple issues.

Most negotiation protocols are evaluated based on one's own testbed. For example, [6] and [7] are only evaluated on randomly generated utility spaces.

T. Ito et al. (Eds.): Innovations in Agent-Based Complex Automated Negotiations, SCI 319, pp. 89–105.
springerlink.com 

However, the effectiveness of the negotiation protocols is evaluated based on the same testbed. Thus, in this paper we propose a tool that generates testbeds for evaluating multi-issue negotiation protocols by focusing on the utility function based on cube-based constraints [10] and cone-constraints. Cone-constraints capture the intuition that agent utilities for a contract usually decrease gradually (rather than step-wise) by the distance from their ideal contract.

We propose a common testbed generating tool based on XML. The input is the configuration files that define the number of issues, the number of agents, etc. The testbed generating tool produces XML files that define the agent's utility spaces in XML format as output. This tool has four types of utility spaces: Random, A Single Hill, Two-Hills, and Several Hills. These types of utility spaces are based on actual negotiation settings.

In this paper, we define XML tags, which represent utility spaces, that consist of cone-based and cube-based constraints. By utilizing an XML format, users can easily understand, modify, and update the meaning of the data and exchange the data among research communities. In addition, our XML format does not depend on a certain environment. In this paper, we show cube-based and cone-based constraint tags that define the building blocks of utility function spaces.

We also demonstrate some examples that use our testbed. First, we show a JAVA program that searches for agreement contracts in agent utility spaces using Simulated Annealing (SA). In this program, the XML structure is analyzed using Document Object Model (DOM)[20], and then agreement points are searched for. Second, we demonstrate experiments that utilize our testbeds for evaluating the Distributed Mediator Protocol (DMP) and Hybrid Secure Protocol (HSP) proposed in [6].

The remainder of the paper is organized as follows. First, we describe a model of nonlinear multi-issue negotiation. Second, we propose a testbed generating tool based on XML for multi interdependent issues. Third, we demonstrate examples using our testbed. Finally, we describe related works and draw a conclusion

## 2 Nonlinear Utility Function

In the literature of multi-issue negotiations, we consider the situation where $n$ agents want to reach an agreement with a mediator who manages the negotiation from the middle position. There are $m$ issues, $s_j \in S$, to be negotiated. The number of issues represents the number of utility space dimensions. For example, if there are three issues, the utility space has three dimensions. The issues are not "distributed" over agents, who are all negotiating a contract with N (e.g., 10) issues in it. All agents are potentially interested in the values for all N issues. Issue $s_j$ has a value drawn from the domain of integers $[0, X]$, *i.e.*, $s_j \in [0, X](1 \leq j \leq M)$. A contract is represented by a vector of issue values $\mathbf{s} = (s_1, ..., s_m)$. The objective function for agreement search protocols can be described as follows:

$$\arg\max_{\mathbf{s}} \sum_{i \in N} u_i(\mathbf{s}).$$

The proposed protocols in the literature try to find contracts that maximize social welfare, *i.e.*, the total utilities for all agents. Such contracts, by definition, will also be Pareto-optimal.

In this paper, we deal with cube-constraints and cone-constraints as the utility function. Every agent has its own, typically unique, set of constraints.

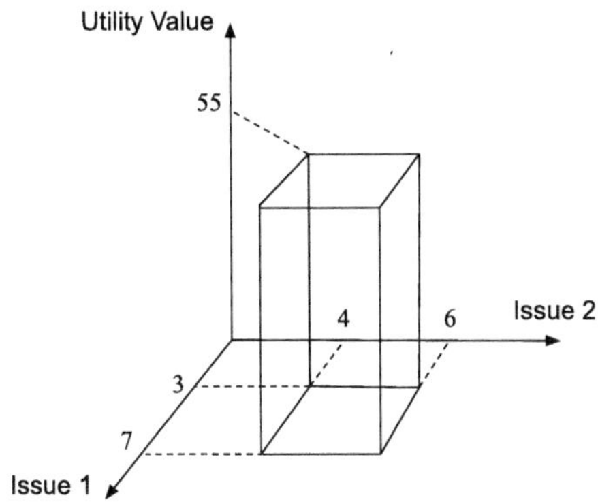


**Fig. 1.** Example of a cube-constraint

**[Cube-constraints]** An agent's utility function is described in terms of constraints [10]. There are $l$ constraints, $c_k \in C$. Each constraint represents a region with one or more dimensions and has an associated utility value. Constraint $c_k$ has value $w_i(c_k, \mathbf{s})$ if and only if it is satisfied by contract $\mathbf{s}(1 \leq k \leq l)$. We call this type of constraint a "cube-constraint." Figure 1 shows an example of a binary constraint between Issues 1 and 2. This constraint, which has a value of 55, holds if the value for Issue 1 is in the range $[3, 7]$ and the value for Issue 2 is in the range $[4, 6]$.

In recent works (e.g., [11]), several types of cube-constraints were proposed. We also include a variety of cube-contraints in our testbed.

**[Cone-constraints]** An agent's utility function can be described in terms of cone-constraints. By formalizing risk attitude in terms of the cone-constraints, utility function of agents capture the utility information in real world. Figure 2 shows an example of a binary cone-constraint between Issues 1 and 2. This cone-constraint has a value of 20, which is maximum if the situation is $\mathbf{s}_{central} = [2, 2]$. The impact region is $\mathbf{w} = [1, 2]$. The expression for a segment of the base is $(x_1 - 2)^2 + (x_2 - 2)^2/4 = 1^1$.

Suppose there are $l$ cone-constraints, $C = \{c_k | \ 1 \leq k \leq l\}$. Cone-constraint $c_k$ has gradient function $g_k(\mathbf{s}_{central}, \mathbf{w})$, which is defined by two values: central value $\mathbf{s}_{central}$, which is the highest utility in $c_k$, and impact region $\mathbf{w}$, which represents the region where $c_k$ is affected. We assume not only circle-based but also ellipse-based cones. Thus constraint $c_k$ has value $u_i(c_k, \mathbf{s})$ if and only if it is satisfied by

[1] The general expression is $\sum_{i=1}^{m} x_i^2 / w_i^2 = 1$.

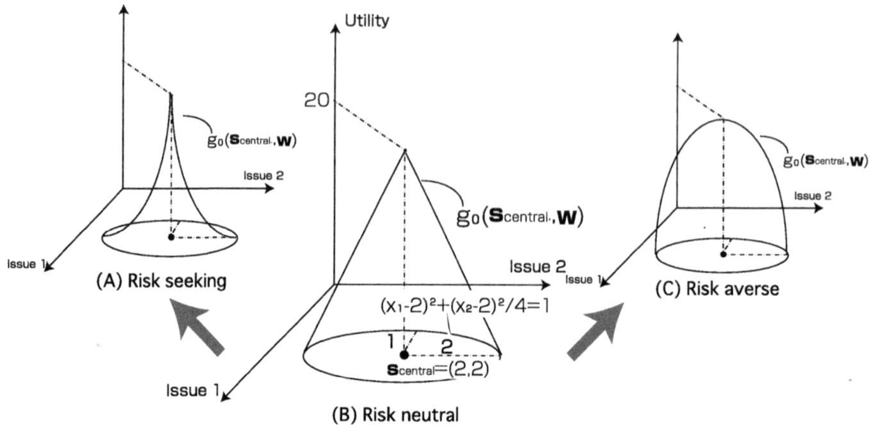


**Fig. 2.** Example of cone-constraints

contract s. In this paper, impact region **w** is not a value but a vector. These formulas can represent utility spaces if they are in a $n$-dimensional space.

In addition, cone-constraints can include the risk attitude for constraints by configuring gradient function $g_k(\mathbf{s}_{central}, \mathbf{w})$. If the agent usually has a risk neutral attitude for $c_k$, $g_k$ is defined as (B) in Fig. 2 (e.g., proportion). However, the attitudes (types) of agent can change from risk-seeking to risk-averse for making agreements. For example, if agents have a risk-seeking attitude for constraint $c_k$, $g_k$ is defined as (A) in Fig. 2 (e.g., exponent). If an agent has a risk-averse attitude for $c_k$, $g_k$ is defined as (C) in Fig. 2. If agents have the most risk-averse attitude for $c_k$, $g_k$ stays constant. Therefore, $c_k$ is shaped like a column if the agents have the most risk-averse attitude. In real world, there are at least two kinds of risk: 1) the risk of getting a bad deal 2) the risk of failing to get a deal. In this paper, we assume " 2) risk of failing to get a deal".

An agent's utility for contract s is defined as $u_i(\mathbf{s}) = \sum_{c_k \in C, \mathbf{s} \in x(c_k)} w_i(c_k, \mathbf{s})$, where $x(c_k)$ is a set of possible contracts (solutions) of $c_k$. This expression produces a "bumpy" nonlinear utility space with high points where many constraints are satisfied and lower regions where few or no constraints are satisfied.

Figure 3 shows an example of a nonlinear utility space with two issues. This utility space is highly nonlinear with many hills and valleys. [10] proposed a utility function based on "cube"-constraints. Compared with cube-constraints, highest point in the utility space is narrower. Therefore, the protocols for making agreements must search in highly nonlinear utility space. A simple simulated annealing method to directly find optimal contracts is especially insufficient in a utility function based on cone-constraints.

We assume, as is common in negotiation contexts, that agents do not share their utility functions with each other to preserve a competitive edge. Generally, in fact, agents do not completely know their desirable contracts in advance, because their

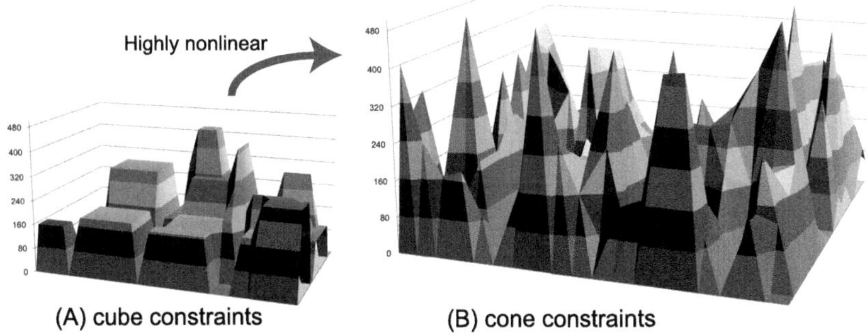


**Fig. 3.** Example of utility space with cone constraints

own utility functions are simply too large. If we have 10 issues with 10 possible values per issue, for example, this produces a space of $10^{10}$ (10 billion) possible contracts, which is too many to evaluate exhaustively. Agents must thus operate in a highly uncertain environment.

# 3 Common Testbed Based on XML for Negotiation Protocols

## 3.1 Testbed Generating Tool

We have been implementing a common testbed generating tool for multi-issue negotiation protocols based on XML. The input of a testbed generating tool is a configuration file that includes the number of issues and the number of agents. The output is an XML file that defines the agents' utility spaces. The source code for this tool is downloadable from: `http://www-itolab.mta.nitech.ac.jp/MultiIssueNegotiations`.

Figure 4 shows the program flow of our testbed generating tool. First, the utility space is defined based on the configuration file. Second, constraints are generated based on the specified type of utility spaces. Finally, an XML file is outputted. The details of the testbed generating tool are shown as follows:

**(1) Defining utility space:** The testbed generating tool defines the utility space information based on the configuration file. The configuration file includes the number of issues, agents, and constraints as well as the value domain per issue. Constraints are classified by the number of related constraints. For example, a unary constraint is related to one issue, a binary constraint is related to two issues, etc. In the configuration file, we write the number of constraints for each related constraint like "unary constraints include 10, binary constraints include 5, etc."

**(2) Generating utility spaces:** In the current implementation, the testbed generating tool generates utility spaces based on four different types of utility spaces:

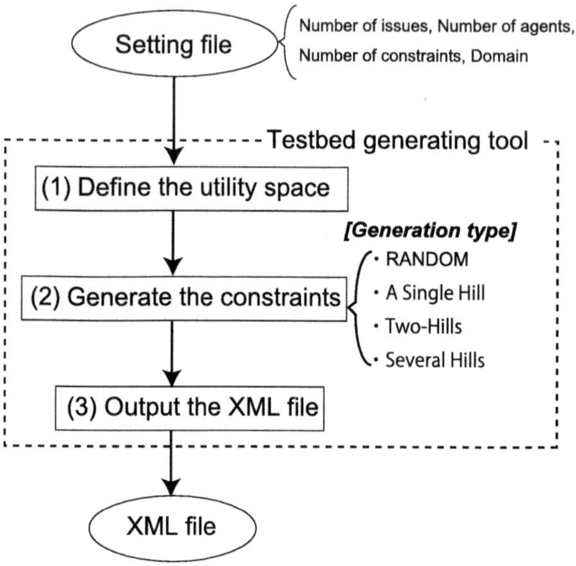


**Fig. 4.** Flow of testbed generating tool

Random, A Single Hill, Two Hills, and Several Hills. Statements about the details of each type are shown as follows:

Random: In this type, constraints are generated randomly. Such generation is used in the experiments in several works [10]. Figure 5 shows an example of utility space plotted by all statements as agent constraints. This utility space plotted is highly nonlinear, as Figure 5(A) shows.

A Single Hill: An example of this type is a collaborative negotiation among the same type of agents. The utility space plotted by all agents has one higher point, as Figure 5(B) shows. In such utility spaces, reaching an agreement is usually easy.

Two Hills: An example of this type is a bilateral negotiation between two types of agents. In particular, such negotiation between buyers and sellers is popular. The utility space plotted by all agents has two higher points, as Figure 5(C) shows. In such utility spaces, making agreements is hard because the agents are likely in a hostile relation.

Several Hills: An example of this type is collaborative negotiation among more than three other types of agents. Collaborative design for a car among designers, engineers, and business managers is a concrete example. The utility space plotted by all agents' constraints has more than three higher points, as Figure 5(C) shows. In such utility spaces, finding agreement points is hard because there are too many hills. Thus search algorithms usually try to find the highest points.

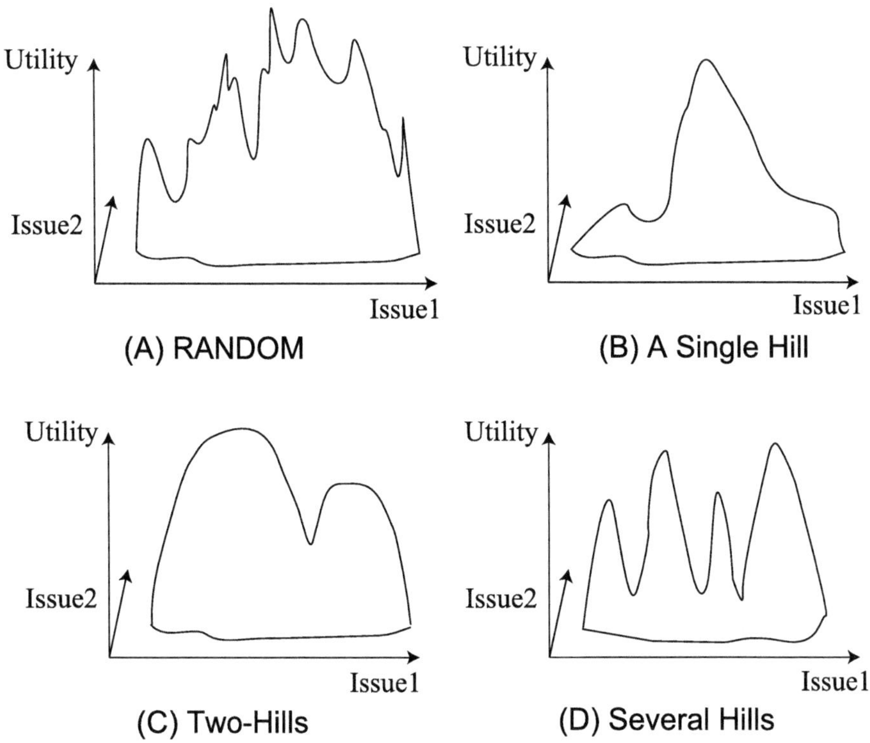


**Fig. 5.** Generation type

**(3) Output XML file:** The testbed generating tool outputs the XML file on the testbed for negotiation. By outputting these files, users can easily understand the information. Additionally, users can modify, change, and update the data, and XML data are not dependent on a certain environment. Users like research communities can also easily exchange data with each other. The details of the XML tags are described in the next subsection.

### 3.2 XML Format for Testbeds

We propose the XML format for expressing the agent's utility function. In XML, this information is defined by tags. The specification of XML tags in cube-constraints and cone-constraints is described as follows:

**XML format for cube-constraints:** Figure 6 shows an example of the XML format for cube-constraints. Figure 7 shows a tree-structured chart for cube-constraints. The tree-structured chart enables us to understand the parent-child relation between tags. A detailed description of the tags is described as follows:

<UtilitySpace>: <Utility Space> tag shows the specification information about the entire utility space. This tag has the elements of <Dimension>, <ValueNumber>, and <Agent>.

<Dimension>: This tag specifies the number of issues. In Figure 6, the number of issues is four.

<Domain>: This tag specifies the value domain for each issue. In Figure 6, the domain of all issues is [0,9].

<Agent>: This tag, which specifies the agents, has attributes of agent's id and name. In Figure 6, the agent's id is 0 and its name is Alice. There could

```
<?xml version="1.0" encoding="Shift_JIS" standalone="no"?>
<UtilitySpace>
<Dimension>4</Dimension>
<Domain>0-9</Domain>
<Agent no=0 name="Alice">
<ReservationValue>11</ReservationValue>
<Constraint no=0 name="0">
<Cardinality>2</Cardinality>
<Utility>69</Utility>
<Minimum>
<Issue no=2 name="size"> 4 </Issue>
</Minimum>
<Maximum>
<Issue no=2 name="size"> 8 </Issue>
</Maximum>
</Constraint>
<Constraint no=1 name="1">
...
</Constraint>
</Agent>
</UtilitySpace>
<Agent no=1 name="Bob">
<ReservationValue>15</ReservationValue>
<Constraint no=0 name="0">
<Cardinality>1</Cardinality>
...
</Constraint>
</Agent>
</UtilitySpace>
```

**Fig. 6.** Example XML for cube-constraints

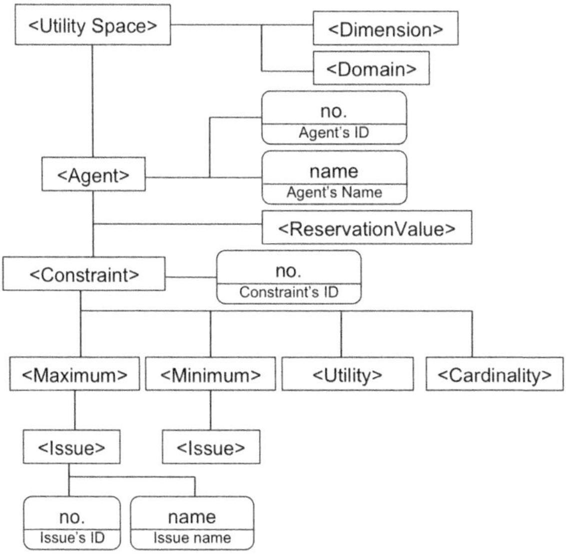


**Fig. 7.** Tree-structured XML chart for cube-constraints

be multiple agent tags in <UtilitySpace> tag. This tag has the elements of <ReservationValue> and many <Constraint> tags.

<ReservationValue>: This tag specifies the reservation utility value for determining whether to "agree" or "disagree" with the contract alternatives in a negotiation. In Figure 6, the reservation value is 21.

<Constraint>: This tag, which defines the constraints, has the id of the constraint as an attribute. This tag has the elements of <Issue>, <Utility>, and <Cardinality>. In Figure 6, the id of the constraints is 0.

<Minimum>: This tag defines the possible minimum values for each issue. In Figure 6, the possible minimum value of Issue 2 is 4. This means that the value for the issue should have more than 4.

<Maximum>: This tag defines the possible maximum values for each issue. In Figure 6, the possible maximum value of Issue 2 is 8. This means that the value for the issue should have less than 8.

<Utility>: This tag defines the utility value in this constraint. The constraints have this utility value if the value for each issue is in the range defined by <Issue> tags. In Figure 6, constraint 0 has a value of 69, and it holds if the value for Issue 1 is 0, the value for issue 2 is 8, the value for Issue 3 is in the range $[4, 8]$, and the value for Issue 4 is 4.

<Cardinality>: This tag shows the number of issues related to this constraint. In Figure 6, the cardinality is one. This is because this constraint is related to issue 2. In the other words, this constraint is constrained by a issues. In our definition, the contract has a value if only the issues related to the constraints

```
<?xml version="1.0" encoding="Shift_JIS" standalone="no"?>
<UtilitySpace>
<Dimension>5</Dimension>
<Domain>0-10</Domain>
<Agent name="Alice" no="0">
<ReservationValue>21</ReservationValue>
<Constraint no="0">
<Cardinality>1</Cardinality>
<MaxUtility>122</MaxUtility>
<RiskAttitude>1</RiskAttitude>
<CenterPoint>
<Issue name="4" no="4">0</Issue>
</CenterPoint>
<Width>
<Issue name="4" no="4">2</Issue>
</Width>
</Constraint>
<Constraint no="1">
<Cardinality>1</Cardinality>
...
</Constraint>
</Agent>
</UtilitySpace>
```


**Fig. 8.** Cone-constraints XML

satisfy the possible values. In other words, all values are permitted in other issues not related to the constraint.

**XML formats for cone-constraints:** Figure 8 shows an example of an XML for cone-constraints. Figure 9 shows a tree-structured chart for cone-constraints. The XML tags in the <UtilitySpace> and <Agents> tags are almost the same as the XML tags for cube-constraints. A detailed description of the tags in the cone-based constraints is described as follows:

<MaxUtility>: This tag shows the central value, which is the highest utility in the constraint. In Figure 8, the central value is 122, which is the maximum utility in the constraint.

<RiskAttitude>: This tag shows a gradient function that represents the risk attitude for making agreements. In our testbed generating tool, we defined a gradient function for each number. For example, one is defined that a gradient function constant is constant. In Figure 8, the risk attitude for making agreements is

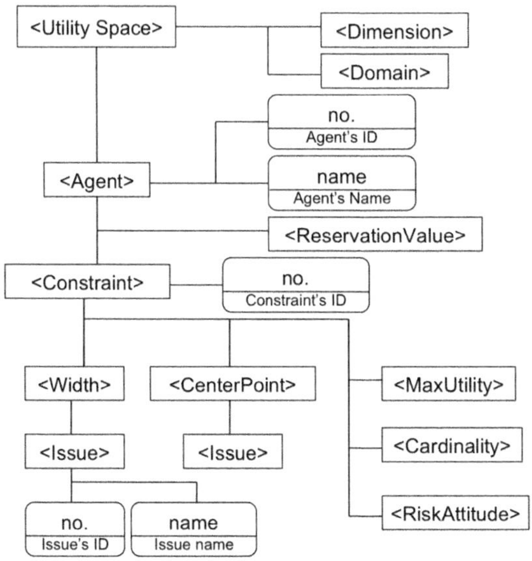


**Fig. 9.** Tree-structured chart for cone-constraints XML

one. Future work includes an extension that enables users to simply define the gradient function.

<Width>: This tag shows the impact region, which represents the region affected by the constraint. The impact region is defined in each <Issue> tag. In Figure 8, the impact region in Issue 4 is two.

<CenterPoint>: This tag shows the central point, where the utility is maximum. In the <CenterPoint> tag, the central point is defined by <Issue> tags. In Figure 8, the central point is 0 in Issue 4 and all values are permitted in other issues (Issues 0 - 3).

# 4 Examples with Testbed

## 4.1 Java Program Using the Testbed

In this subsection, we describe the Java program using the testbeds proposed in the previous section. Our code was implemented in Java 2 (1.5). The program source codes are downloadable from: `http://www-itolab.mta.nitech.ac.jp/MultiIssueNegotiations/`.

Figure 10 shows the flow of the JAVA program using testbeds. This program inputs XML files generated by the tool. The following are the details of this program behavior:

Analyzing XML files: In this program, an XML file is analyzed by a Document Object Model (DOM)[20], which is a platform and a language-independent standard object model for representing HTML or XML documents as well as an

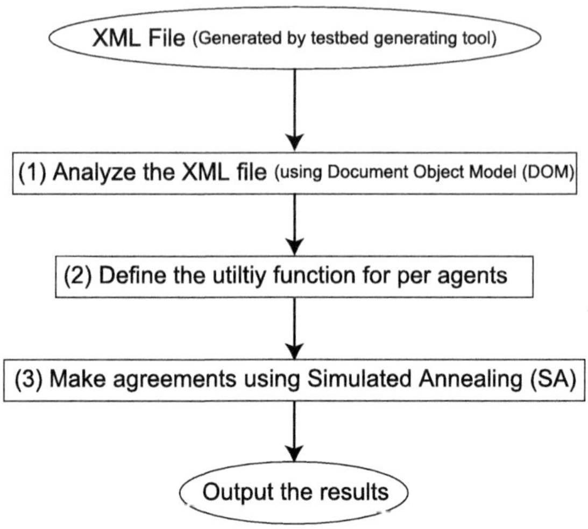


**Fig. 10.** Program flow using testbeds

Application Programming Interface (API) for querying, traversing, and manipulating such documents. The information of the structure of the utility space and the agent's utility function are read from XML files.

Defining the utility function for each agent: The structure of the utility space and the agent's utility function are defined based on the XML analyzed in the previous step.

Searching agreements using SA: In this program, we provide a simple agreement algorithm that gathers and aggregates all individual agent's utility spaces into one central place and then finds the most optimal contract using simulated annealing (SA) [18]. In simulated annealing, the mediator moves randomly if the temperature is high, but he/she moves to the highest neighbor if the temperature is low. A simulated-annealing method of making agreements was employed in previous works [10] because this search method is superior to other search methods, such as hill climbing search in multi interdependent issue negotiation.

In future work, we will generate this program using other programming languages such as C++, Ruby, Python, and Perl so that this testbed can be used by many users.

### 4.2 Example of Experiments for Evaluating the Negotiation Protocols

We demonstrate some experimental results to show that our past proposed protocols can utilize our testbed. In the experiments, we show the experimental results of the Distributed Mediator Protocol (DMP) and the Hybrid Secure Protocol (HSP) [6] with the testbed. The details of these protocols are described in [8][6].

In each experiment, we ran 100 negotiations between agents with Random, A Single Hill, Two Hills, and Several Hills. The following are the parameters for our experiments. The number of agents was six, and the number of mediators was four.

We compared the following methods: "(A) DMP (SA)" is the Distributed Mediator Protocol and the search algorithm is simulated-annealing. "(B) DMP (HC)" is the Distributed Mediator Protocol and the search algorithm is hill-climbing. "(C) DMP (GA)" is the Distributed Mediator Protocol and the search algorithm is the genetic algorithm. "(D) HSP (SA)" is the hybrid secure protocol, and the search algorithm in the distributed mediator step is simulated annealing. "(E) HSP (HC)" is the hybrid secure protocol, and the search algorithm in the distributed mediator step is the hill-climbing algorithm.

These experiments show the optimality rate in five protocols. The optimality rate is defined as *(maximum utility value calculated by each method) / (DMP (SA))*. Since DMP (SA) usually finds a high optimal contract, we consider it the basis for comparing other methods.

**Utility function:** All constraints are one-constraints. The domain for the issue values is $[0, 9]$. Constraints include $10$ unary constraints, $5$ binary constraints, and $5$ trinary constraints, etc. (A unary constraint is related to one issue, a binary constraint is related to two issues, and so on). The value in the central point is *100* $\times$ *(Number of issues)*. The maximum impact region for a constraint is 7.

The gradient function is defined as four types.

Type 0: $u(\mathrm{s}) = (Max\ Value) * log(e - (distance)/(width) * (e-1))$.
Type 1: $u(\mathrm{s}) = (Max\ Value) * (1 - (distance)/(width))$.
Type 2: $u(\mathrm{s}) = (1 - (Max\ Value))^{(distance)/(width)}$ + *(Max Value)-1*.
Type 3:$u(\mathrm{s}) = (Max\ Value)$.
($u(\mathrm{s})$: utility value at s when $sisinthecone - constraints$, $(distance)$: distance between s and the central point, $(width)$: impact region, *(Max Value)*: value at the central point).

We set the following parameters for the search methods: HC, SA, and GA.

**Hill climbing (HC):** The number of iterations is *20* + *(Number of issues)* $\times$ *5*. The final result is the maximum value achieved.

**Simulated annealing (SA):** The initial temperature is 50. For each iteration, the temperature is decreased by 0.1. Thus, it decreased to 0 by 500 iterations. *20* + *(Number of issues)* $\times$ *5* searches are conducted while the initial start point is being changed.

**Genetic algorithm (GA):** The population size in one generation is *20* + *(Number of Issues)* $\times$ *5*. We employed a basic crossover method in which two parent individuals are combined to produce two children (one-point crossover). The fitness function is the sum of all agents' (declared) utility. 500 iterations were conducted. Mutations happened at very small probability. In a mutation, one of the issues in a contract vector was randomly chosen and changed. In the GA-based method, we define an individual as a contract vector.

Our code was implemented in Java 2 (1.5) and run on a core 2-duo processor iMac with 1.0 GB memory on a Mac OS X 10.5 operating system.

Figures 11 $\sim$ 14 show the optimality rate in five methods. In all types of generation, "(C) DMP (GA)" is the highest of the five methods, not only RANDOM.

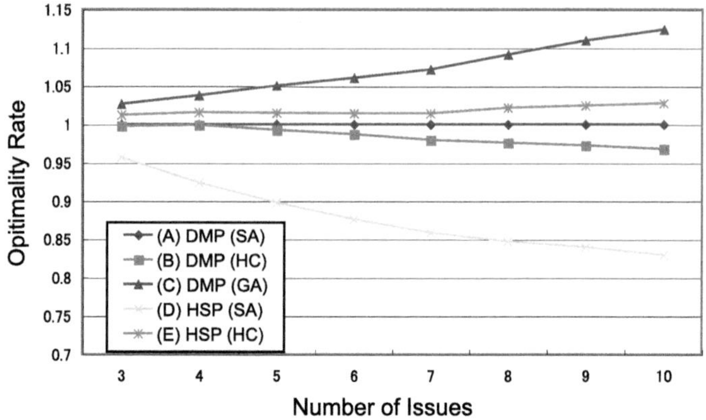


**Fig. 11.** Optimality rate (RANDOM)

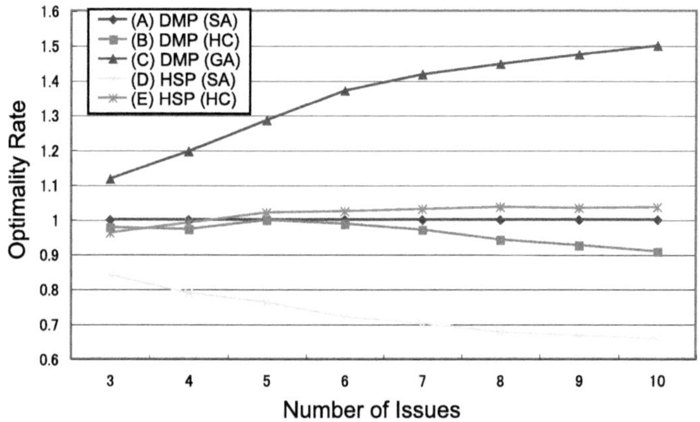


**Fig. 12.** Optimality rate (A Single Hill)

Therefore, "(C) DMP (GA)" has high quality to find the optimal contract. Meanwhile, "(B) DMP (HC)" decreases rapidly based on the number of issues in all types of generation, because hill climbing reaches local optima by increasing the search space. "(E) HSP (HC)" slightly outperforms "(A) DMP (SA)" in all types of generation. Therefore, DMP (GA) and HSP (HC) are better for finding the optimal agreement point in all types of generating, not only in RANDOM.

Comparing Figures 11 12, "(C) DMP (GA)" in Figure 12 has a higher value than "(C) DMP (GA)" in 11. "(D) HSP (SA)" in Figure 12 also has a lower value than "(D) HSP (SA)" in 11. Therefore, the difference of the optimality rates among the

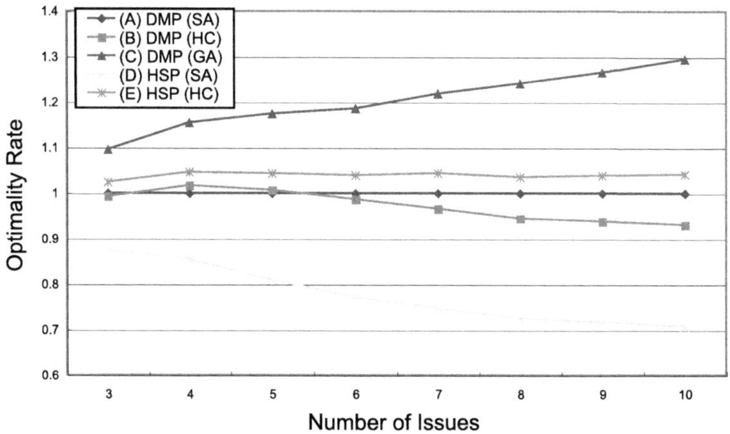


**Fig. 13.** Optimality rate (Two-Hills)

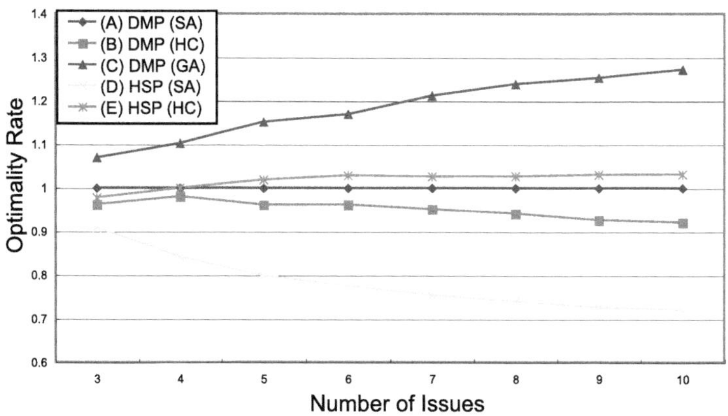


**Fig. 14.** Optimality rate (Several Hills)

methods in "A SIngle Hill" is larger than one in "RANDOM" because the value in the local optima in "A Single Hill" is a lower value than one in random.

## 5 Related Works

As far as the authors know, this is the first attempt to create a testbed for multiple interdependent issue negotiation protocols. The following is a literature review of multi-issue negotiation problems. All of these protocols are evaluated on the original testbed. Our testbed might provide opportunities to compare these algorithms based on the same criteria.

Most previous work on multi-issue negotiation ([2, 3, 4]) has only addressed linear utilities. Some researchers have been focusing on more complex and nonlinear utilities.

[15] explored a range of protocols based on mutation and selection on binary contracts. This paper does not describe what kind of utility function is used, nor does it present any experimental analyses, so it remains unclear whether this strategy enables sufficient exploration of utility space. [1] presents an approach based on constraint relaxation.

[12] presented a protocol that was applied with near-optimal results on medium-sized bilateral negotiations with binary dependencies. The work presented here is distinguished by demonstrating both scalability and high optimality values for multilateral negotiations and higher order dependencies.

[13] and [14] also presented a protocol for multi-issue problems for bilateral negotiations. [25] and [16] presented a multi-item and multi-issue negotiation protocol for bilateral negotiations in electronic commerce situations.

[5] proposed bilateral multi-issue negotiations with time constraints, and [19] proposed multi-issue negotiations that employ a third party to act as a mediator to guide agents toward equitable solutions. This framework also employs an agenda that serves as a schedule for the ordering of issue negotiations. Agendas are very interesting because agents only need to focus on a few issues. [9] proposed a checking procedure to mitigate this risk and showed that by tuning this procedure's parameters, outcome deviation can be controlled. These studies reflect interesting viewpoints, but they focused on just bilateral trading or negotiations.

## 6 Conclusion

In this paper, we proposed a testbed generating tool based on XML for multi-issue negotiation. Our tool provides a common testbed to evaluate the effectiveness of multi-issue negotiation protocols. Moreover, users can easily understand the meaning of data because it is based on a simple XML format. In this testbed, four types of utility spaces were provided that corresponded to real negotiation cases. Finally, we demonstrated examples of experiments using our testbed in which we analyzed the differences among types of utility spaces.

## References

1. Barbuceanu, M., Lo, W.-K.: Multi-attribute utility theoretic negotiation for electronic commerce. In: Dignum, F.P.M., Cortés, U. (eds.) AMEC 2000. LNCS (LNAI), vol. 2003, pp. 15–30. Springer, Heidelberg (2001)
2. Bosse, T., Jonker, C.M.: Human vs. computer behaviour in multi-issue negotiation. In: Proc. of 1st International Workshop on Rational, Robust, and Secure Negotiations in Multi-Agent Systems (RRS 2005), pp. 11–24 (2005)
3. Faratin, P., Sierra, C., Jennings, N.R.: Using similarity criteria to make issue trade-offs in automated negotiations. Artificial Intelligence 142, 205–237 (2002)

4. Fatima, S.S., Wooldridge, M., Jennings, N.R.: Optimal negotiation of multiple issues in incomplete information settings. In: Proc. of Third International Joint Conference on Autonomous Agent and Multi-Agent Systems (AAMAS 2004), pp. 1080–1087 (2004)
5. Fatima, S.S., Wooldridge, M., Jennings, N.R.: Approximate and online multi-issue negotiation. In: Proc. of the 6th Inernational Joint Conference on Autonomous Agents and Multi-agent Systems (AAMAS 2007), pp. 947–954 (2007)
6. Fujita, K., Ito, T., Klein, M.: Preliminary result on secure protocols for multiple issue negotiation problems. In: Bui, T.D., Ho, T.V., Ha, Q.T. (eds.) PRIMA 2008. LNCS (LNAI), vol. 5357, pp. 161–172. Springer, Heidelberg (2008)
7. Fujita, K., Ito, T., Klein, M.: A representative-based multi-round protocol for multi-issue negotiations. In: Proc. of The 7th Inernational Joint Conference on Autonomous Agents and Multi-agent Systems, AAMAS 2008 (2008)
8. Fujita, K., Ito, T., Klein, M.: Approximately fair and secure protocols for multiple interdependent issues negotiation. In: Proc. of The 8th Inernational Joint Conference on Autonomous Agents and Multi-agent Systems (AAMAS 2009) (2009)
9. Hindriks, K., Jonker, C., Tykhonov, D.: Avoiding approximation errors in multi-issue negotiation with issue dependencies. In: Proc. of The 1st International Workshop on Agent-based Complex Automated Negotiations (ACAN 2008) (2008)
10. Ito, T., Hattori, H., Klein, M.: Multi-issue negotiation protocol for agents: Exploring nonlinear utility spaces. In: Proc. of 20th International Joint Conference on Artificial Intelligence (IJCAI 2007), pp. 1347–1352 (2007)
11. Ivan Marsa-Maestre, J.R.V., Lopez-Carmona, M.A., de la Hoz, E.: Effective bidding and deal identification for negotiations in highly nonlinear scenarios. In: Proc. of The 8th Inernational Joint Conference on Autonomous Agents and Multi-agent Systems (AAMAS 2009) (2009)
12. Klein, M., Faratin, P., Sayama, H., Bar-Yam, Y.: Negotiating complex contracts. Group Decision and Negotiation 12(2), 58–73 (2003)
13. Lai, G., Li, C., Sycara, K.: A general model for pareto optimal multi-attribute negotiations. In: Proc. of The 2nd International Workshop on Rational, Robust, and Secure Negotiations in Multi-Agent Systems (RRS 2006) (2006)
14. Lai, G., Sycara, K., Li, C.: A decentralized model for multi-attribute negotiations with incomplete information and general utility functions. In: Proc. of The 2nd International Workshop on Rational, Robust, and Secure Negotiations in Multi-Agent Systems (RRS 2006) (2006)
15. Lin, R.J., Chou, S.T.: Bilateral multi-issue negotiations in a dynamic environment. In: Proc. of AMEC 2003 (2003)
16. Robu, V., Poutre, H.L.: Retrieving the structure of utility graphs used in multi-item negotiation through collaborative filtering of aggregate buyer preferences. In: Proc. of The 2nd International Workshop on Rational, Robust, and Secure Negotiations in Multi-Agent Systems (RRS 2006) (2006)
17. Robu, V., Somefun, D.J.A., Poutre, J.L.: Modeling complex multi-issue negotiations using utility graphs. In: Proc. of the 4th International Joint Conference on Autonomous Agents and Multi-Agent Systems (AAMAS 2005) (2005)
18. Russell, S.J., Norvig, P.: Artificial Intelligence: A Modern Approach. Prentice Hall, Englewood Cliffs (2002)
19. Shew, J., Larson, K.: The blind leading the blind: A third-party model for bilateral multi-issue negotiations under incomplete information. In: Proc. of The 1st International Workshop on Agent-based Complex Automated Negotiations (ACAN 2008) (2008)
20. W3C: Document object model (dom) technical reports (2010), `http://www.w3.org/DOM/DOMTR`

# Constraint and Bid Quality Factor for Bidding and Deal Identification in Complex Automated Negotiations

Ivan Marsa-Maestre[1], Miguel A. Lopez-Carmona[1], Takayuki Ito[2], Mark Klein[2], Juan R. Velasco[1], and Katsuhide Fujita[3]

[1] Computer Engineering Department, Universidad de Alcala, Spain
{ivan.marsa,miguelangel.lopez,juanramon.velasco}@uah.es
[2] Center for Collective Intelligence, Massachusetts Institute of Technology, USA
{takayuki,m_klein}@mit.edu
[3] School of Techno-Business Administration, Nagoya Institute of Technology, Japan
juanramon.velasco@uah.es

**Summary.** Complex automated negotiations usually involve multiple, interdependent issues. These negotiation scenarios are specially challenging because the agents' utility functions are nonlinear, which makes traditional negotiation mechanisms not applicable. Even mechanisms designed and proven useful for nonlinear utility spaces may fail if the utility space is highly nonlinear. For example, although both contract sampling and constraint sampling have been successfully used in auction based negotiations with constraint-based utility spaces, they tend to fail in highly nonlinear utility scenarios. In this paper, we will show that the performance of these approaches decrease drastically in these negotiation scenarios, and propose a mechanism which balances utility and deal probability for the bidding and deal identification processes. The experiments show that the proposed mechanisms yield better results than the previous approaches in terms of optimality and scalability.

## 1 Introduction

Integrative negotiation approaches intend to allow negotiating agents to search for joint gains when pursuing an agreement [16]. In the last years, there has been an increasing interest in complex negotiations scenarios where agents negotiate about multiple, interdependent issues [13]. These scenarios are specially challenging, since issue interdependency yields nonlinear utility functions for the agents, and thus the classic mechanisms for linear negotiation models are not applicable. A more in-depth review on utility space complexity and related-research on automated negotiations involving complex utility spaces is provided in Section 2.

In particular, this work focuses on multilateral mediated negotiation, where several agents try to reach an agreement over a range of issues using a bidding based negotiation protocol with the aid of a mediator. The utility spaces for the agents are

T. Ito et al. (Eds.): Innovations in Agent-Based Complex Automated Negotiations, SCI 319, pp. 107–128.
springerlink.com 

generated using weighted constraints, which results in nonlinear utility functions. In [11], a bidding mechanism is proposed, which is based on taking random samples of the contract space and applying simulated annealing to these samples to identify high utility regions for each agent, sending these regions as bids to a mediator, and then performing a search in the mediator to find overlaps between the bids of the different agents. In a similar scenario [20], samples are taken from the constraints space instead. Experiments show that these approaches achieve high effectiveness (measured as high optimality rates and low failure rates for the negotiations) in the evaluation scenario they describe (Section 3). However, as we will show empirically in Section 6.2, these approaches perform worse as the circumstances of the scenario turn harder (that is, when the utility functions are highly nonlinear, like B2B interactions or distributed automated control systems). Under these circumstances, the failure rate increases drastically, raising the need for an alternative approach.

Furthermore, as described in [11], the bidding-based negotiation protocol presents some scalability concerns due to the extensive search for overlaps performed in the mediator, which finally limits the maximum number of bids each agent may send depending on the number of agents in the negotiation. In this paper, we intend to address these problems in the following ways:

- We propose a mechanism to take into account both the utility of a bid for an agent and its *viability* (a measure of the likelihood of the bid to yield a deal), and integrate this mechanism in the contract sampling and constraint sampling approaches (Section 4). We will show that this balance between bid utility and deal probability yields a significant improvement in terms of optimality rate and failure rate over the previous approaches in highly nonlinear scenarios.
- We propose a heuristic search mechanism for the mediator which lowers the scalability problem while achieving acceptable optimality rates (Section 5).

A highly-nonlinear simulated scenario has been devised to validate our hypotheses and evaluate the effects of our contributions. This scenario is described in Section 6, along with the discussion of the results obtained. Finally, our proposal is briefly compared to the most closely-related works in the state-of-the-art (Section 7). The last section summarizes our conclusions and sheds light on some future research.

## 2 Complex Utility Spaces

In the last years, there has been an increasing interest in complex negotiation scenarios, where agents negotiate about multiple, interdependent issues [13]. These scenarios are specially challenging, since issue interdependency yields nonlinear utility spaces, which make classic negotiation approaches not applicable [11]. In this context, it is necessary to be able to measure the complexity of a given utility space, in order to assess whether the studied scenario is beyond the capabilities of traditional negotiation mechanisms. If this is the case, alternative negotiation mechanisms, specifically designed for complex environments, will be needed. In this section we first briefly discuss utility space complexity and present the most widely used metrics for the assessment of the complexity of a solution space, and then we review the most relevant works so far on the field of automated negotiation in complex utility spaces.

### 2.1 Utility Space Complexity

Characterization of the structural complexity of utility spaces in negotiation is an analogous problem to function characterization for optimization. In this context, and more specifically in the field of optimization using evolutionary algorithms, structural complexity analysis plays a crucial role, since algorithm search capabilities are greatly impacted by some structural properties of the optimized function, which is usually known as *fitness landscape* in evolutionary computation.

A fitness landscape $L$ may be defined as a tuple [9]

$$L = \langle S, f, F, >_F, \phi \rangle ,$$

where:

- $S$ is the solution space (or *genotype space*, as it is called in evolutionary computation).
- $f : S \rightarrow F$ is a fitness function, analogous to the utility functions which are used in negotiation.
- $F$ represents the range of values for the fitness functions. For the case of utility functions, it is usual that $F \subset \mathbb{R}$.
- $>_F$ represents a partial ordering in $F$, which allows to compare two solutions $v, w \in S$ given the values of $f(v)$ and $f(w)$. $f(v) >_F f(w)$ means that solution $v$ is preferred against $w$.
- $\phi$ is a *neighborhood* operator, defined as $\phi : S \times S \rightarrow [0, 1]$, such that given $v, w \in S$, $\phi(u, v)$ represents the probability that the search function (usually, a genetic algorithm) passes from point $u$ to point $w$.

The $\phi$ function is directly related to the search mechanism used and its parameters (e.g. simulated annealing temperature or mutation probability for genetic algorithms), which implies an important consequence: the complexity of a utility space may be different depending on the considered search algorithm and its parameters. This operator also defines the concept of *neighbor solutions* in the space. Two solutions $v, w \in S$ will be neighbors if $\phi(u, v)$ is over a given threshold, and will be denoted $v \in N(w)$ or $w \in N(n)$. In many fitness landscapes in the literature, $\phi$ is a binary operator.

The structural properties of a fitness landscape which are interesting regarding search complexity within the space are determined by the properties of the local optima in the landscape. A solution $s \in S$ is a local maximum if its fitness value is greater or equal than the fitness values of *all* its neighbors, that is, $f(s) \geq f(v) \, \forall v \in N(s)$. Local minima are defined in an analogous manner.

$Modality$ is defined as the number of local optima in the space. Despite the intuitive notion, modality is not directly related to complexity. An unimodal space may be very complex for a genetic algorithm if its gradient structure is not adequate. On the other hand, a maximally multimodal space, where half of the solutions are local optima but there is a clear, monotonic structure, is very simple for evolutive search algorithms [10].

*Ruggedness* is often confused in the literature with modality. However, ruggedness in a highly modal landscape does not refer to the number of local optima, but to the distribution of these optima throughout the landscape, that is, to the variety of forms like peaks, valleys and ridges. Depending on this distribution, landscape ruggedness may or may not imply high complexity. Generally, landscapes with structured ruggedness distributions will be less complex than those presenting noisy ruggedness.

For low number of local optima, we can have *smooth* or *neutral* utility landscapes [27]. Landscape smoothness is defined taking into account the existence of *attraction basins* towards its local optima. An attraction basin towards a solution $s_n$ is defined as the set of solutions

$$B(s_n) = \{s_0 \in S \,|\, \exists s_1, \ldots, s_n \in S \text{ con } s_{i+1} \in N(s_i) \text{ y } f(s_{i+1}) >_F f(s_i) \,\forall i\}.$$

The size of the basin is given by the cardinality of the set $B(s_n)$. The larger the attraction basins in a landscape, the smoother the landscape.

The opposite property to smoothness is *neutrality*, caused by the existence of plateaus and ridges in the landscape. Two solutions $s_1$ and $s_2$ are neutral if $f(s_1) = f(s_2)$. A *plateau* is defined as the set of solutions

$$P(s_n) = \{s_0 \in S \,|\, \exists s_1, \ldots, s_n \in S \text{ con } s_{i+1} \in N(s_i) \text{ y } f(s_{i+1}) = f(s_i) \,\forall i\}.$$

In the evolutionary computation literature we find also the concept of *deceptive landscapes*. From an intuitive point of view, a landscape is deceptive if it tends to lead the search algorithm to solutions other than the optimal [7]. The main problem of this concept is that it depends strongly on the search algorithm used. For instance, a highly modal utility space with a high number of local optima with lower utility than the global optimum would be deceptive for a direct search algorithm (it tends to get stuck in local optima), but not for simulated annealing or genetic algorithms. This dependence on the search method makes deceptiveness not useful to compare different approaches.

Once the properties which has an influence on the complexity of a fitness landscape or a solution space have been described, techniques which allow to measure the complexity of a given space are needed. One of the approaches we can find in the literature is to study the relationship between the value of the fitness function $f$ in a given point of the solution space and the distance to that point distance from the global optimum. In most cases, a high correlation between the distance to a point from the global optimum and the fitness value for this point (*fitness distance correlation, FDC*) should make the search for optima easier, so this concept may be used to predict the difficulty of the negotiation or optimization problem. Evaluation may be performed quantitatively, by means of the classic expression for the correlation coefficient for a set of samples

$$r = \frac{C_{FD}}{\sigma_F \sigma_D},$$

where

$$C_{FD} = \frac{1}{n}\sum_{i=1}^{n}\left(f_1 - \overline{f}\right)\left(d_1 - \overline{d}\right)$$

is the covariance of the values of the fitness function $f_i$ and the distances from the global optimum $d_i$ for the samples considered, and $\overline{f}$, $\overline{d}$, $\sigma_F$ y $\sigma_D$ are, respectively, the means and standard deviations. However, quantitative evaluation often does not reflect accurately landscape structure, specially for anisotropic solution spaces. In these cases, a qualitative evaluation may be performed using scatter plots [26]. The main problem with this metric is the need to know the global optimum of the fitness function $f$. An alternative way to measure complexity is the *local fitness-distance correlation*, where the correlation coefficient is computed using the distance to the nearest local optima.

Another statistic which is widely used to assess the complexity of a fitness landscape or solution space is autocorrelation. Autocorrelation computes, for every distance $d$, the correlation between a function and the same function displaced a distance $d$. Computation is made performing a "random walk" throughout the landscape, using the $\phi$ neighborhood operator to determine movement between successive steps in the walk. This metric allows for a qualitative evaluation of the space structure by means of a representation of the autocorrelation function of the random walk. To make quantitative evaluations, *correlation length* is used. Correlation distance is defined as the minimum distance $\psi$ which makes correlation fall below a given threshold (usually 0.5), which gives an idea of the distance we can move throughout the solution space while keeping a certain correlation between samples [18]. As it happened with fitness-distance correlation, correlation length depends on a notion of distance with will generally vary with the search mechanism used, which makes the metric less generic.

The problem with those metrics is that they do not assess adequately anisotropic functions. In [9], an statistic characterization is proposed. It is also based on random walks, but it adds an analysis of the series of samples defined by the walks, which results in an stochastic model representing the correlation structure of the space, instead of generating a single numeric value which summarizes the structure of the whole space. However, it still depends on performing walks throughout the space using the $\phi$ operator, and thus will not be applicable to compare approaches which have different neighborhood operators. In addition, this characterization is computationally harder to perform than the previous ones.

As we can see, there is not a clear, preferred method to assess the complexity of a preference space. In each case, the characteristics of the specific scenario (i.e. anisotropy, neighborhood operator, knowledge about global optima, computational constraints...) must be levered in order to select the most suitable technique to characterize the structural complexity of the utility spaces we are dealing with.

## 2.2 Related Research on Automated Negotiation in Complex Utility Spaces

In the context of a multi-attribute negotiation, complexity of a preference space may depend at least on the number of issues, the level of interdependency between the preferences on the issues, the domain of the issues, the possibility of change over time of the negotiation context, and the method used to describe preferences. In general, a large number of issues with a high interdependency and a large domain contribute to more complex preference spaces. If the negotiation context changes over time, complexity also increases. The method to describe preferences also has an influence in the complexity of the preference space. A constraint-based preference space for instance may present discontinuities which make gradient based optimizers not applicable, while differentiable utility functions contribute to a faster local optimization. In this section we review the most representative research on multi-attribute negotiation for complex preference spaces.

Klein *et al.* [13] present, as far as we are aware, the first negotiation protocols specific for complex preference spaces. They propose a simulated annealing-based approach, a refined version based on a parity-maintaining annealing mediator, and an unmediated version of the negotiation protocol. Of great interest in this work are the positive results about the use of simulated annealing as a way to regulate agent decision making, along with the use of agent expressiveness to allow the mediator to improve its proposals. However, this expressiveness is somewhat limited, with only four possible valuations which allow the mediator to decide which contract to use as a parent for mutation, but not in which direction to mutate it. On the other hand, the performed experiments only consider the bilateral negotiation scenario, though agents claim that the multiparty generalization is simple. Finally, the family of negotiation protocols they propose are specific for binary issues and binary dependencies. Higher-order dependencies and continuous-valued issues, common in many real-world contexts, are known to generate more challenging utility landscapes which are not considered in their work.

Luo *et al.* [17] propose a fuzzy contraint based framework for multi-attribute negotiations. In this framework a buyer agent defines a set of fuzzy constraints to describe its preferences. The proposals of the buyer agent are a set of hard constraints which are extracted from the set of fuzzy constraints. The seller agent responds with an offer or with a relaxation request. The buyer then decides whether to accept or reject an offer, or to relax some constraints by priority from the lowest to highest. In [15, 16] an improvement to Luo's model is presented. They devise an expressive negotiation protocol where proposals include a valuation of the different constraints, and seller's responses may contain explicit relaxation requests. It means that a seller agent may suggest the specific relaxation of one or more constraints. The relaxation suggested by a seller agent is based on utility and viability criteria, which improves the negotiation process. Though these constraint-based works model discontinuous preference spaces, the operators used to compute utility and the utility spaces defined yield monotonic preference spaces, which are far from the complex preference spaces covered in our work.

Another interesting approach to solve the computational cost and complexity of negotiating interdependent issues is to simplify the negotiation space. Hindriks *et al.* [8] propose a weighted approximation technique to simplify the utility space. They show that for smooth utility functions the application of this technique results in an outcome that closely matches the outcome based on the original interdependent utility structure. The method is evaluated for a number of randomly generated utility spaces with interdependent issues. Experiments show that this approach can achieve reasonably good outcomes for utility spaces with simple dependencies. However, an approximation error that deviates negotiation outcomes from the optimal solutions cannot be avoided, and this error may become larger when the approximated utility functions become more complex. Authors acknowledge as a necessary future work to study which kind of functions can be approximated accurately enough using this mechanism. Another limitation of this approach is that it is necessary to estimate a region of utility space where the actual outcome is expected to be (i.e. it is assumed that the region is known a priori by the agents).

In [25] utility graphs are used to model issue interdependencies for binary-valued issues. Utility graphs are inspired by graph theory and probabilistic influence networks to derive efficient heuristics for non-mediated bilateral negotiations about multiple issues. The idea is to decompose highly non-linear utility functions in sub-utilities of clusters of inter-related items. They show how utility graphs can be used to model an opponent's preferences. In this approach agents need prior information about the maximal structure of the utility space to be explored. Authors argue that this prior information could be obtained through a history of past negotiations or the input of domain experts. However, our approach has the advantage that outcomes can be reached without any prior information and that it is not restricted to binary-valued issues.

There are several proposals which employ genetic algorithms to learn opponent's preferences according to the history of the counter-offers based upon stochastic approximation. In [3] a system based on genetic-algorithms for electronic business is proposed. In this work the utility functions are restricted to take a product combination form (i.e. utility of an outcome is the product of the utility values of the different issues). The objective function used is based on the comparison of the changes of consecutive offers. Small changes of an issue suggest that this issue is more important. For each new population, the protocol enforces that the generated candidates cannot be better than the previous offer. Unlike other negotiation models based on genetic algorithms, this proposal adapts to the environment by dynamically modifying its mutation rate. Lau *et al.* [14] have also reported a negotiation mechanism for non-mediated automated negotiations based on genetic algorithms. The fitness function relies on three aspects: an agent's own preference, the distance of a candidate offer to the previous opponent's offer, and time pressure. In this work agents' preferences are quantified by a linear aggregation of the issue valuations. However, non-monotonic and discontinuous preference spaces are not explored. In [4] a genetic algorithm is proposed which is based on a joint elitism operation and a joint fitness operation. In the joint elitism operation an agent stores the latest offers received from the opponent. The joint fitness operation combines agent's own utility function and euclidean distance to the opponent's offer. In this work two different negotiation scenarios are considered. In the first one utility is defined as the weighted sum of the different issue values (i.e. issues are independent). The second scenario defines a utility function where there is a master issue and a set of slave issues. Utility is calculated as the weighted sum of the different issue values, but the weights of the slave and master issues change according to the value of the master issue.

In [28] a mediated negotiation framework for multi-agent negotiation is presented. This framework involves a mediation step in which the individual preference functions are aggregated to obtain a group preference function. The main interest is focused on the implementation of the mediation rule where they allow a linguistic description of the rule using fuzzy logic. A notable feature of their approach is the inclusion of a mechanism rewarding the agents for being open to alternatives other than simply their most preferred. The negotiation space and utility values are assumed to be arbitrary (i.e. preferences can be non-monotonic). However, the set of possible solutions is defined a priori and is fixed. Moreover, the preference function needs to be provided to the mediation step in the negotiation process, and pareto-optimality is not considered. Instead, the stopping rule is considered, which determines when the rounds of mediation stop.

Fatima *et al.* [6] analyze bilateral multi-issue negotiation involving nonlinear utility functions. They consider the case where issues are divisible and there are time constraints in the form of deadlines and discounts. They show that it is possible to reach Pareto-optimal agreements by negotiating all the issues together, and that finding an equilibrium is not computationally easy if the agents' utility functions are nonlinear. In order to overcome this complexity they investigate two solutions: approximating nonlinear utilities with linear ones; and using a simultaneous procedure where the issues are discussed in parallel but independently of each other. This study shows that the equilibrium can be computed in polynomial time. An important part of this work is the complexity analysis and estimated approximation error analysis performed over the proposed approximated equilibrium strategies. Heuristic approaches have generally the drawback of the lack of a solid mathematical structure which guarantees their viability, which raises the need of an exhaustive experimental evaluation. An adequate complexity analysis and establishing a bound over the approximation error contribute to give heuristic approaches part of the technical soundness they usually lack. Among the limitations of the proposal, we can point out that this work is focused on symmetric agents where the preferences are distributed identically, and the utility functions are separable in nonlinear polynomials of a single variable. This somewhat limits the complexity of the preference space.

In summary, in the existing research nearly all the models which assume issue interdependency rely on monotonic utility spaces, binary valued issues, low-order dependencies, or a fixed set of defined a priori solutions. Simplification of the negotiation space has also been reported as a valid approach for simple utility functions, but it cannot be used with higher-order issue dependencies, which generate highly uncorrelated utility spaces. Therefore, new approaches are needed if automated negotiation is to be applied to non-mediated settings involving non-monotonic, highly uncorrelated preference spaces.

# 3 Contract and Constraint Sampling for Negotiation in Complex Utility Spaces

## 3.1 Constraint-Based Nonlinear Utility Spaces

Nonlinear agent preferences can be described by using different categories of functions, like K-additive utility functions [2], bidding languages [24], or weighted constraints [12]. In this work we focus on nonlinear utility spaces generated by means of weighted constraints. In these cases, agents' utility functions are described by defining a set of constraints. Each constraint represents a region with one or more dimensions, and has an associated utility value. The number of dimensions of the space is given by the number of issues $n$ under negotiation, and the number of dimensions of each constraint must be lesser or equal than $n$. The utility yielded by a given potential solution (contract) in the utility space for an agent is the sum of the utility values of all the constraints that are satisfied by that contract. Figure 1 shows an example for two issues and three constraints: a unary constraint $C1$ and two binary constraint $C2$ and $C3$. The utility values associated to the constraints are also shown in the figure. In this example, contract $x$ would yield a utility value for the agent $u(x) = 15$, since it satisfies both $C1$ and $C2$, while contract $y$ would yield a utility value $u(y) = 5$, because it only satisfies $C1$. It can also be noted that unary constraint $C1$ can be seen as a binary constraint where the width of the constraint for issue 2 is all the domain of the issue, so we can generalize and say that all constraints have $n$ dimensions.

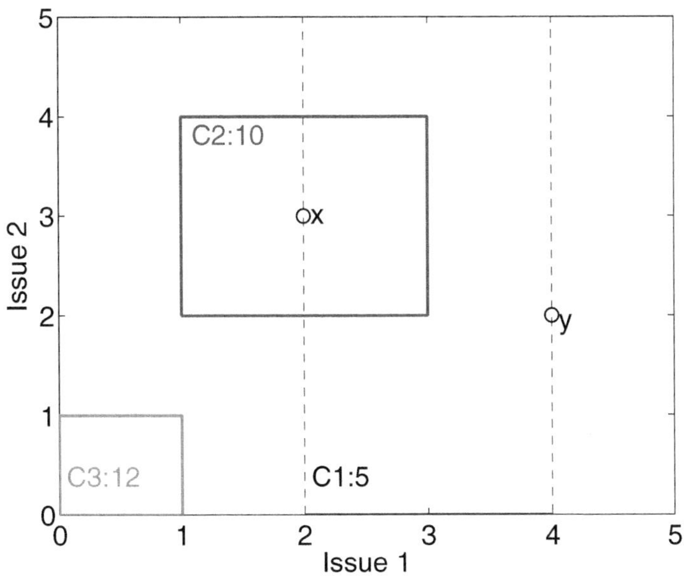


**Fig. 1.** Example of a utility space with two issues and three constraints.

More formally, we can define the issues under negotiation as a finite set of variables $x = \{x_i | i = 1, ..., n\}$, and a contract (or a possible solution to the negotiation problem) as a vector $s = \{x_i^s | i = 1, ..., n\}$ defined by the issues' values. Issues take values from the domain of integers $[0, X]$.

Agent utility space is defined as a set of constraints $C = \{c_k | k = 1, ..., l\}$. Each constraint is given by a set of intervals which define the region where a contract must be contained to satisfy the constraint. In this way a constraint $c$ is defined as $c = \{I_i^c | i = 1, ..., n\}$, where $I_i^c = [x_i^{min}, x_i^{max}]$ defines the minimum and maximum values for each issue to satisfy the constraint. Constraints defined in this way describe hyper-rectangular regions in the $n$-dimensional space. Each constraint $c_k$ has an associated utility value $u(c_k)$.

A contract $s$ satisfies a constraint $c$ if and only if $x_i^s \in I_i^c \forall i$. For notation simplicity, we denote this as $s \in x(c_k)$, meaning that $s$ is in the set of contracts that satisfy $c_k$. An agent's utility for a contract $s$ is defined as $u(s) = \sum_{c_k \in C | s \in x(c_k)} u(c_k)$, that is, the sum of the utility values of all constraints satisfied by $s$. This kind of utility functions produces nonlinear utility spaces, with high points where many constraints are satisfied, and lower regions where few or no constraints are satisfied.

### 3.2 Contracts Sampling and Simulated Annealing in Bidding-Based Nonlinear Negotiation

In our previous work [11], we presented a bidding-based protocol to deal with nonlinear utility spaces generated using weighted constraints. The protocol consists of the following four steps:

1. *Sampling:* Each agent takes a fixed number of random samples from the contract space, using a uniform distribution.
2. *Adjusting:* Each agent applies simulated annealing to each sample to try to find a local optimum in its neighborhood. This results in a set of high-utility contracts.
3. *Bidding:* Each agent generates a bid for each high-utility, adjusted contract. The bids are generated as the intersection of all constraints which are satisfied by the contract. Bids defined in this way represent hyperrectangle regions in the $n$-dimensional solution space. Each agent sends its bids to the mediator, along with the utility associated to each bid.
4. *Deal identification:* The mediator employs breadth-first search with branch cutting to find overlaps between the bids of the different agents. The regions of the contract space corresponding to the intersections of at least one bid of each agent are tagged as potential solutions. The final solution is the one that maximizes joint utility.

The protocol is evaluated in a nonlinear scenario for different number of agents and issues, and it achieves good results in terms of optimality (measured as the ratio between the solutions found using the protocol and the optimal solution computed using complete information) and failure rate (measured as the ratio between unsuccessful negotiations and total negotiations).

### 3.3 Maximum Weight Independent Set and the Max-Product Algorithm

In [20], we proposed an alternative perspective for the bidding process, based on looking at the constraint-based agent utility space as a weighted undirected graph. Consider again the simple utility space example shown in Figure 1. Think about each constraint as a node in the graph, with an associated weight which is the utility value associated to the constraint. Now we will connect all nodes whose corresponding constraints are *incompatibles*, that is, they have no intersection. The resulting graph is shown in Figure 2.

To find the highest utility bid in such a graph can be seen as finding the set of unconnected nodes which maximizes the sum of the nodes' weights. Since only incompatible nodes are connected, the corresponding constraints will have non-null intersection. In the example, this would be achieved by taking the set $\{C1, C2\}$. The problem of finding a maximum weight set of unconnected nodes is a well-known problem called maximum weight independent set (MWIS). Though MWIS problems are NP-hard, in [1], a message passing algorithm is used to estimate MWIS. The algorithm is a reformulation of the classical max-product algorithm called "min-sum", and works as follows. Initially, every nodes $i$ send their weights $\omega_i$ to their neighbors $N(i)$ as messages. At each iteration, each node $i$ updates the message to send to each neighbor $j$ by subtracting from its weight $\omega_i$ the sum of the messages received from *all other* neighbors *except* $j$. If the result is negative, a zero value is sent as message. Upon receiving the messages, a node is included in the estimation of the $MWIS$ if and only if its weight is greater than the sum of al messages received from its neighbors. Message passing continues until $MWIS$ converges or the maximum number of iterations is exceeded. This is formally shown in Algorithm 1.

**Algorithm 1.** Min-sum algorithm for MWIS estimation

**Input**:
$i = 1,...,n$: nodes (constraints) in the weighted graph $\omega_i | i = 1,...,n$: weight (utility) of each node (constraint) $N(i)$: set of neighbors of each node (incompatible constraints)
$t_{max}$: maximum number of iterations
**Output**: $MWIS$: estimation of the MWIS
$t = 0; m^t_{i\to j} = \omega_i \forall j \in N(i)$
**while** $t < t_{max}$ **do**
  $t = t+1$; **foreach** $i$ **do**
    $m^t_{i\to j} = \max\{0, \omega_i - \sum_{k\neq j, k\in N(i)} m^{t-1}_{k\to i}\}$
  **end**
  $MWIS^t = \{i | \omega_i > \sum_{k\in N(i)} m^{t-1}_{k\to i}\}$
  **if** $t > 1$ *and* $MWIS^t = MWIS^{t-1}$ **then**
    return $MWIS^t$
**end**

Since the algorithm is deterministic, only one bid can be generated for a given set of constraints. To solve this, in [20], the algorithm is applied to a subset of constraints $C' = \{c'_k | k = 1, ..., n_c; n_c < l; c'_k \in C\}$. The constraints $c'_k$ are randomly chosen from the constraint set $C$. In this way, a different constraint subset $C'$ is passed to the algorithm at each run, which will result in different, non-deterministic bids.

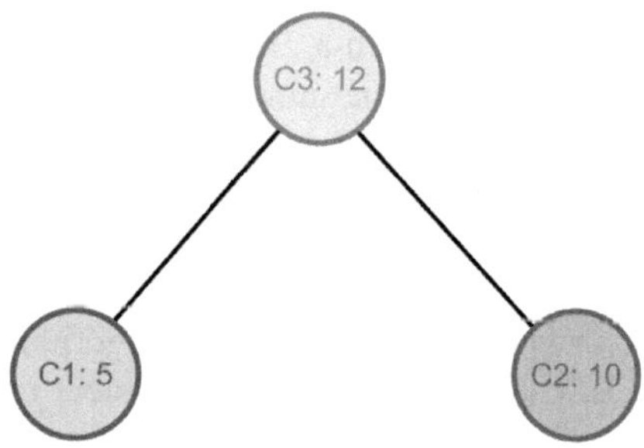


**Fig. 2.** Weighted undirected graph resulting from the utility space in Figure 1.

## 4 Bidding Mechanisms for Highly-Nonlinear Utility Spaces

The use of weighted constraints generates a "bumpy" utility space, with many peaks and valleys. However, the degree of "bumpiness" is highly dependent on the way the constraint set is generated, and specially on the average width of the constraints. In [11], constraints are generated by choosing the width of each constraint in each issue randomly within the [3,7] interval. Since the domain is chosen to be [0,9], this generates rather "wide" constraints. Figure 3 shows an example of the resulting two-dimensional utility space for 50 "wide" binary constraints. On the other hand, Figure 4 shows an utility space obtained using "narrow" constraints. Comparing both figures we can see that, though both utility spaces are nonlinear, the space generated using narrow constraints is more complex, with narrower peaks and valleys. As the number of issues under consideration increases, the differences between having wide or narrow constraints become more relevant. Though the approaches proposed in [11] and [20] work perfectly in scenarios like the example shown in Figure 3, we will see that their performance (in terms of optimality and failure rate) decreases drastically in highly nonlinear scenarios defined using narrow constraints, and therefore an alternative approach is needed to deal with these highly nonlinear utility spaces.

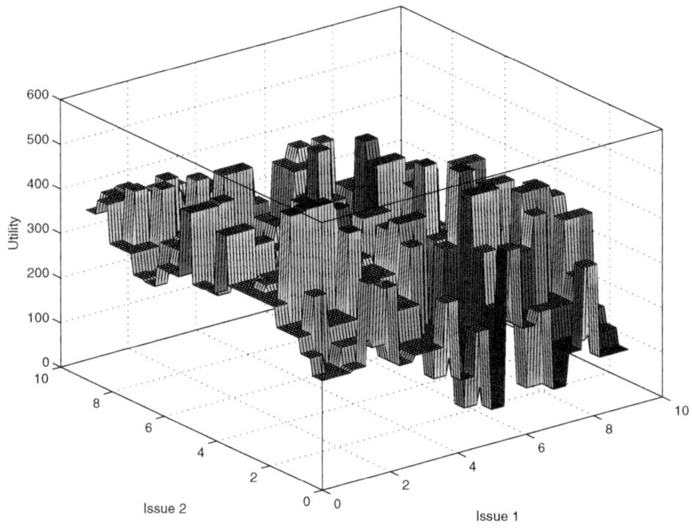


**Fig. 3.** Example of a nonlinear utility space generated by using "wide" constraints.

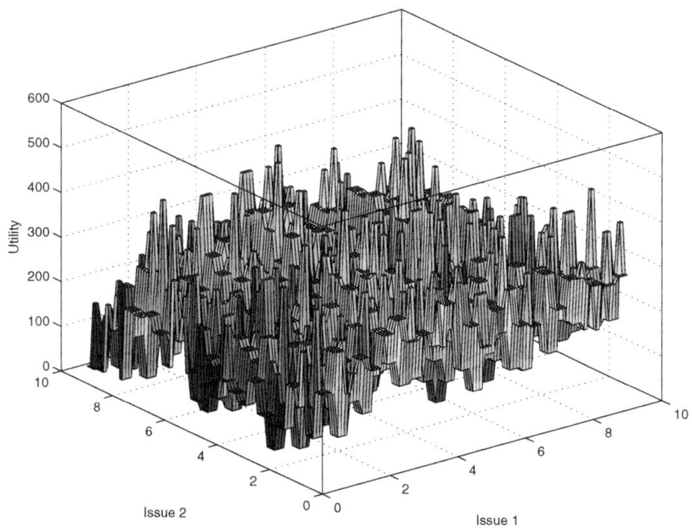


**Fig. 4.** Example of a highly nonlinear utility space generated by using "narrow" constraints.

## 4.1 Constraint/Bid Quality Factor

If we compare the utility spaces shown in Figures 3 and 4, we can see that the main difference between them (apart from the absolute utility values, but they have no effect in optimality) is the width of the peaks. Highly-nonlinear scenarios will yield narrower peaks. Since the mechanisms outlined above lead agents to choose those peaks (or high-utility regions) as bids, the result is that narrower bids will be sent to the mediator. The width of the bids (or more generally, the volume of the bids in the $n$-dimensional space), will directly impact the probability that the bid overlaps a bid of another agent, and thus the probability of the bid resulting in a deal. Intuitively, an agent with no knowledge of the other agents' preferences should try to adequately balance the utility of their bids (to maximize its own profit) and the volume of those bids (to maximize the probability of a successful negotiation). To formally represent this, we define the *quality factor* of a constraint or a bid as $Q_c = u_c^{\alpha} \cdot v_c^{1-\alpha}$, where $u_c$ and $v_c$ are, respectively, the utility and volume of the bid or constraint $c$, and $\alpha \in [0, 1]$ is a parameter which models the risk attitude of the agent. A risk averse agent ($\alpha < 0.5$) will tend to qualify as better bids those that are wider, and thus are more likely to result in a deal. A risk willing or selfish agent ($\alpha > 0.5$) will, in contrast, give more importance to bid utility.

Our hypothesis is that by taking into account this quality factor in the bidding mechanisms, with adequate values for the parameter $\alpha$, will result in a better balance between utility and "width" in agent bids, and thus negotiations will yield higher optimality rates and lower failure rates.

### 4.2 Using the Quality Factor within the Simulated Annealing Algorithm

To make the simulated annealing bidding approach to take advantage of the quality factor Q is fairly straightforward. We just need to make the simulated annealing optimizer to search for contracts which maximize the quality factor Q instead of the agent utility. Since the quality factor Q is a feature of a region, not a contract, the adjusted contracts must be mapped to the high utility regions where they are contained before they are accepted or rejected by the simulated annealing engine. This can be easily done by checking all constraints in the agent preference model and computing the intersection of the constraints which are satisfied by the candidate contract. The volume of this intersection can then be used to compute the quality factor Q of the region. The mechanism can be seen in Algorithm 2. The function $adjust\_annealing(x, Q(., \alpha), n_{SA}, T_{SA})$ uses simulated annealing to returns a region of optimal quality factor using as starting point a sampled contract $x$ (1).

### 4.3 Q-Based Tournament Selection for the MWIS Approach

The quality factor Q cannot be directly introduced into the max-product or min-sum algorithm, because the algorithm is based in a weighted graph where weights are additive, and the quality factor is not additive (that is, the quality factor of the intersection of a set of constraints is not the sum of the quality factor of the constraints). Thus, a different approach is needed to introduce this factor in the algorithm. The

**Algorithm 2.** Bid generation using simulated annealing over quality factor

**Input**:
- $D$: solution space domain
- $n_b$: maximum number of bids
- $u_R$: reservation utility for the agent
- $C$: constraint set defining agent's utility space
- $u$: agent's utility function
- $\alpha$: agent's attitude parameter
- $Q$: function which computes the quality factor of a region
- $n_{SA}$: iteration bound for the simulated annealing algorithm
- $T_{SA}$: initial temperature for the simulated annealing algorithm

**Output**:
- $B$: bid set

$B = \varnothing$;
$k = 0$;
**while** $k < n_b$ **do**
  $k = k + 1$;
  $x = random\_contract()$;
1   $b = adjust\_annealing(x, Q(., \alpha), n_{SA}, T_{SA})$;
  **if** $u(b) \geq u_R$ **then**
    $B = B \bigcup b$;
**end**
$remove\_duplicates(B)$

**Algorithm 3.** Bid generation using MWIS and Q-based tournament selection

**Input**:
- $n_b$: maximum number of bids
- $u_R$: reservation utility for the agent
- $C$: constraint set defining agent's utility space
- $\Omega$: constraint weights for the agents
- $u$: agent's utility function
- $n_c$: number of randomly chosen constraints passed to the MWIS algorithm
- $n_{MWIS}$: maximum number of iterations for the MWIS algorithm
- $\alpha$: agent's attitude parameter
- $n_t$: number of candidate subsets in tournament selection

**Output**:
- $B$: bid set

$B = \varnothing$;
$k = 0$;
**while** $k < n_b$ **do**
  $k = k + 1$;
1   $C' = tournament\_selection(C, n_c, \alpha, n_t)$;
  $\{nodes, weights, neighbors\} = build\_tree(C', \Omega)$;
  $MWIS = minsum(nodes, weights, neighbors, n_{MWIS})$;
  $b = generate\_bid(C', MWIS)$;
  **if** $u(b) \geq u_R$ **then**
    $B = B \bigcup b$;
**end**
$remove\_duplicates(B)$

approach proposed in [21] can be seen in Algorithm 3. We propose to use a *tournament selection* [22] based on the constraint quality factor $Q$ when generating the subset of constraints $C'$ to be passed to the max-product algorithm (1). This tournament selection works as follows. For each bid to generate, a number $n_t$ of candidate constraint subsets are randomly generated. From these subsets, the one which maximizes the product of the quality factors Q of its constraints is chosen as the subset $C'$ to be used for the max-product algorithm. In this way, since high-$Q$ constraints are more likely to be selected, we expect the average Q for the resulting bids to be higher.

## 5 A Probabilistic Mechanism for Deal-Identification

Scalability was identified as one of the main drawbacks in our bidding based negotiation protocol [11]. Once agents have placed their bids, the mediator performs an exhaustive search for overlaps between the bids using a breadth-first algorithm with branch cutting. In a worst case scenario, this means searching through a total of $n_b^{n_a}$ bid combinations, where $n_b$ is the number of bids per agent, and $n_a$ is the number of negotiating agents. In the experiments, we limited the number of combinations to $6,400,000$. This means that, for 4 negotiating agents, the maximum number of bids per agent is $\sqrt[n_a]{6400000} = 50$. This limit becomes harder as the number of agents increases. For example, for 10 agents, the limit is 4 bids per agent, which drastically reduces the probability of reaching a deal. This is specially true for highly-nonlinear utility spaces, where the bids are narrower.

To address this scalability limitation, in [21] we proposed to perform a probabilistic search in the mediator instead of an exhaustive search. This means that the mediator will try a certain number $n_{bc}$ of randomly chosen bid combinations, where $n_{bc} < n_b^{n_a}$. In this way, $n_{bc}$ acts as a performance parameter in the mediator, which limits the computational cost of the deal identification phase. Of course, restricting the search for solutions to a limited number of combinations may cause the mediator to miss good deals. Taking this into account, the random selection of combinations is biased to maximize the probability of finding a good deal. Again, the parameter used to bias the random selection is $Q$, so that higher-$Q$ bids have more probability of being selected for bid combinations at the mediator.

The mechanism is formally shown in Algorithm 4. We can see that the number of analyzed bid combinations is limited to $n_{bc}$ (1), and that the function $combine\_bids$ $(B, U, Q)$ select the bid combinations to analyze (2). Limiting bid combinations at the mediator allows us to remove the limit on the bids issued by the agents, which increases the probability of finding potential deals. Finally, the algorithm selects from all deals found the one which maximizes social welfare, computed using the $sw\,(s, U)$ function (3). In [21], social welfare is computed as the Nash product [23], that is, the product of the utilities that a potential solution gives to every agent.

**Algorithm 4.** Probabilistic deal identification

```
Input:
    A: set of negotiating agents
    n_a = |A|: number of negotiating agents
    B: set of bids issued by every agent
    U: declared utilities for every agent's bids
    Q: declared quality factors for every agent's bids
    sw: social welfare function
    n_bc: maximum number of bid combinations at the mediator
Output:
    s_f: final deal
n = 0;
S = ∅;
while n < n_bc do
    s = combine_bids(A, n_a, B, U, Q);
    if s ≠ ∅ then
        S = S ∪ s;
    n = n + 1;
end
s_f = arg{max_{s∈S} sw(s, U)};
```

# 6 Experimental Evaluation

The hypotheses of this work are that the proposed mechanisms provide an improvement to the optimality of the negotiation process over the previous works described in Section 3. To evaluate this, we have performed a set of experiments to compare the results of the basic approaches with the results obtained introducing the quality factor Q in the bidding and deal identification mechanisms.

## 6.1 Experimental Settings

Several experiments have been conducted to validate our hypotheses. In each experiment, we ran 100 negotiations between agents with randomly generated utility functions. Each negotiation was run for each of the different approaches analyzed. For each set of utility functions we applied a nonlinear optimizer to the product of all agents' utility functions to find the optimal contract and its associated joint utility value. This optimal contract was used to assess the optimality of the different approaches.

We ran experiments with the following parameters:

- Number of agents $n_a = \{4,...,14\}$. Number of issues $n = \{4,...,20\}$. Domain for issue values: integers within the interval $[0,9]$.
- $l$ uniformly distributed random generated constraints per agent: 5 unary constraints, 5 binary contraints, 5 trinary constraints, etc.
- Utility for each $m$-ary constraint drawn from a uniform distribution in the domain $[0, 100 \times m]$.
- Different average widths for constraints, ranging from 2 to 7.

- Settings for simulated annealing: initial tempreature $T_0 = 30$. Number of iterations: 30.
- Maximum number of bids generated per agent $n_b = 200 \times n$.
- Parameters for $Q$ calculation: $\alpha = 0.5$.
- Number of candidate sets in tournament selection $n_t = 10$.
- Number of constraints in each candidate set for tournament selection $n_c = \min(20, l/2)$.
- Maximum number of bid combinations at the mediator: $n_{bc} = 6400000$. For the non-probabilistic, basic mediator, this is achieved by limiting the number of bids sent to the mediator by each agent to $\sqrt[n_a]{6400000}$.
- Joint utility for a failed negotiation: 0.

Experiments were coded in MATLAB and run on a 2x3.2Ghz Qad-Core Intel Xeon processor with 4Gb memory under Mac OS X 10.5.4.

### 6.2 Experimental Results

Figure 5 shows the results of 100 runs of the experiments for 6 agents and 6 issues. The vertical axis represents the median optimality rates of the experiments, while the horizontal axis represents the degree of non-linearity of the utility spaces of the agents, measured using a ruggedness factor based on correlation distance [18]. Four sets of values have been represented:

- basic contract sampling with simulated annealing, and basic constraint sampling with MWIS, represented in dashed lines, with triangle and square vertices, respectively (both lines coincide in the figure).

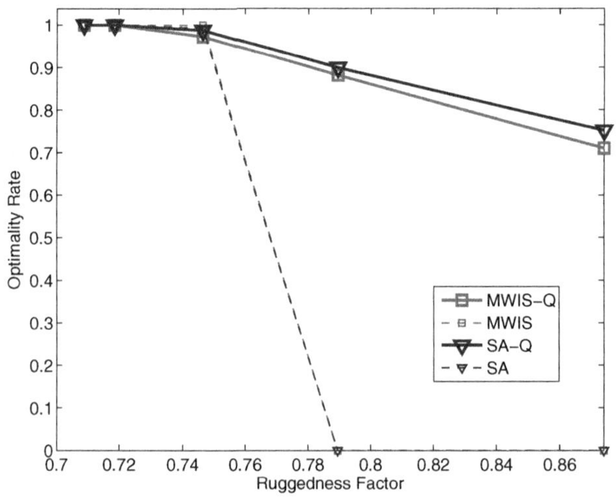


**Fig. 5.** Optimality rate results for 6 agents and 6 issues with different constraint widths.

- contract sampling using Q-based simulated annealing and Q-based mediation, represented in a solid line, with triangle vertices,
- Q-based tournament selection for constraint sampling, MWIS, and Q-based mediation, represented in a solid line, with square vertices.

We can see that both basic contract sampling and basic constraint sampling yield high optimality rates for medium ruggedness, but the median optimality rate decreases drastically (in fact, it drops to zero) as the ruggedness increases (that is, for highly nonlinear utility spaces). The Q-based approaches yield slightly lower optimality rates for wider constraints, which is reasonable, since the Q is used to make a trade-off between utility and deal probability. However, as the agents' preference model turns highly nonlinear, introducing the quality factor Q in the bidding and deal identification mechanisms significantly outperforms the previous approaches, yielding acceptable optimality rates even with the narrowest constraint widths. From these results we can conclude that the quality factor $Q$ can be used to improve failure rate in highly-nonlinear utility spaces, and both simulated annealing and tournament constraint selection with MWIS are suitable ways to select which constraints to use for bid generation.

Regarding scalability, Table 1 shows the optimality rates obtained for one of the studied approaches (MWIS) for a fixed maximum constraint width (4) when the number of agents and issues increases, comparing the results obtained using the basic approach with those obtained introducing the quality factor. We can see that introducing the quality factor Q in the mechanisms significantly improves scalability with the number of agents and issues.

Finally, Table 2 shows the medians and their 95% confidence intervals for the ratio between the bidding, deal identification and total times of our proposed approaches and their corresponding basic approaches. We can see that for both contracts and constraints sampling the use of the quality factor Q introduces a slight overhead over the bidding time, but this overhead is compensated by the significant improvement in deal identification times due to the use of the probabilistic mediator. Also, a comparison of the total negotiation time for both Q-based approaches has been included in the table.

**Table 1.** Scalability with the number of agents and issues

| | | $n$ | | | |
|---|---|---|---|---|---|
| $n_a$ | Approach | 6 | 10 | 14 | 20 |
| 6 | MWIS | 0.9915 | 0.7521 | 0 | 0 |
| | MWIS-Q | 0.9728 | 0.9352 | 0.8903 | 0.8471 |
| 10 | MWIS | 0.6732 | 0 | 0 | 0 |
| | MWIS-Q | 0.9584 | 0.9057 | 0.8551 | 0.7513 |
| 14 | MWIS | 0 | 0 | 0 | 0 |
| | MWIS-Q | 0.8203 | 0.7745 | 0.7232 | 0.5403 |

**Table 2.** Performance comparison

| Approaches | Bidding Time Ratio | |
|---|---|---|
| | median | conf. interval |
| SA-Q / SA | 1.0982 | [1.0888, 1.1076] |
| MWIS-Q / MWIS | 1.0892 | [1.0832, 1.0951] |
| Approaches | Deal Identification Time Ratio | |
| | median | conf. interval |
| Q-mediator / mediator | 0.2018 | [0.1942, 0.2093] |
| Approaches | Total Negotiation Time Ratio | |
| | median | conf. interval |
| SA-Q / SA | 0.8417 | [0.7963, 0.8870] |
| MWIS-Q / MWIS | 0.4285 | [0.4012, 0.4559] |
| MWIS-Q / SA-Q | 0.1792 | [0.1788, 0.1797] |

## 7 Discussion

In [11], the authors propose a single-shot, auction-based protocol which samples the contracts space and uses simulated annealing to identify high utility regions in the agent's utility spaces to be sent as bids to a mediator. In [20], instead of performing a direct sampling of the contract space, different techniques are used over the constraint space to generate bids. We use these works as a starting point to provide effective bidding and deal identification mechanisms for highly-nonlinear utility spaces, where the "narrowness" of the agents' high-utility regions makes the failure rate of their approaches drastically higher. Our approach is based on using a quality factor $Q$, which balances bid utility and bid volume to take into account the likelihood of the bid resulting in a deal. This is a somewhat similar approach to the notion of *viability* seen in [15] for fuzzy-constraint based negotiation or the similarity criteria used in [5] for linear utility spaces.

Other technique for addressing non-linearity in negotiation is to approximate the utility functions by means of linear regression techniques or average weighting methods, as proposed in [8]. However, as authors acknowledge, these approaches are not useful for non-smooth utility spaces.

Finally, there are other works which suggest the use of expressive negotiation protocols in multi-agent negotiations. In [19], gradient information is used to bias the search for solutions in linear unmediated negotiation, and [16] uses relax requirements in bilateral buyer-seller negotiations.

## 8 Conclusions and Future Work

The performance of existing auction-based approaches for negotiation in nonlinear scenarios dramatically decreases when confronted with highly nonlinear scenarios where the negotiating agents' high utility regions are very "narrow" and so it is very unlikely that high utility bids overlap. This paper presents a mechanism to balance bid "width" and bid utility, and integrate this mechanism into two previous approaches. The experiments show that the proposed mechanisms significantly improve the previous approaches in highly nonlinear utility spaces in terms of failure rate and optimality. However, there is still plenty of research to be done in this area. The impact of the parameter $\alpha$ in the optimality rate should be analyzed. In addition, we are interested on designing and evaluating different tournament selection and probabilistic deal identification mechanisms, using different probability density functions. Finally, we are working on iterative negotiation protocols, where agents may change their attitudes or relax their bids as the protocol iterates.

*Acknowledgement.* This work has been partially developed in the framework of the ITEA-2 project 2008005, "Do-it-Yourself Smart Experiences", and partially supported by the Spanish Ministry of Education and Science grant TIN2008-06739-C04-04.

## References

1. Bayati, M., Shah, D., Sharma, M.: Max-product for maximum weight matching: Convergence, correctness, and lp duality. IEEE Transactions on Information Theory 54(3), 1241–1251 (2008)
2. Chevaleyre, Y., Endriss, U., Estivie, S., Maudet, N.: Multiagent resource allocation with k-additive utility functions. In: Proceedings of the DIMACS-LAMSADE Workshop on Computer Science and Decision Theory, pp. 83–100 (2004)
3. Choi, S.P.M., Liu, J., Chan, S.P.: A genetic agent-based negotiation system. Computer Networks 37(2), 195–204 (2001), Electronic Business Systems
4. Chou, T.C., Fu, L.C., Liu, K.P.: E-negotiation of dependent multiple issues by using a joint search strategy. In: ICRA 2007: IEEE International Conference on Robotics and Automation, pp. 1298–1303 (2007)
5. Faratin, P., Sierra, C., Jennings, N.: Using similarity criteria to make negotiation trade-offs. In: Proceedings of the 4th International Conference on Multi-Agent Systems, pp. 119–126 (2000)
6. Fatima, S., Wooldridge, M., Jennings, N.R.: An analysis of feasible solutions for multi-issue negotiation involving nonlinear utility functions. In: AAMAS 2009: Proceedings of The 8th International Conference on Autonomous Agents and Multiagent Systems. International Foundation for Autonomous Agents and Multiagent Systems, Richland, SC, pp. 1041–1048 (2009)
7. Goldberg, D.E., Goldberg, D.E., Deb, K., Horn, J.: Massive multimodality, deception, and genetic algorithms (1992)
8. Hindriks, K., Jonker, C., Tykhonov, D.: Eliminating interdependencies between issues for multi-issue negotiation. In: Klusch, M., Rovatsos, M., Payne, T.R. (eds.) CIA 2006. LNCS (LNAI), vol. 4149, pp. 301–316. Springer, Heidelberg (2006)
9. Hordijk, W.: A measure of landscapes. Working Papers 95-05-049, Santa Fe Institute (1995), `http://ideas.repec.org/p/wop/safiwp/95-05-049.html`
10. Horn, J., Goldberg, D.E.: Genetic algorithm difficulty and the modality of fitness landscapes. In: Foundations of Genetic Algorithms, vol. 3, pp. 243–269. Morgan Kaufmann, San Francisco (1994)
11. Ito, T., Klein, M., Hattori, H.: Multi-issue negotiation protocol for agents: Exploring nonlinear utility spaces. In: Proceedings of the 20th International Joint Conference on Artificial Intelligence (IJCAI 2007), pp. 1347–1352 (2007)

12. Ito, T., Klein, M., Hattori, H.: A multi-issue negotiation protocol among agents with nonlinear utility functions. Journal of Multiagent and Grid Systems 4(1), 67–83 (2008)
13. Klein, M., Faratin, P., Sayama, H., Bar-Yam, Y.: Protocols for negotiating complex contracts. IEEE Intelligent Systems 18(6), 32–38 (2003)
14. Lau, R.Y., Tang, M., Wong, O.: Towards genetically optimised responsive negotiation agents. In: Society, I.C. (ed.) Proceedings of the IEEE/WIC/ACM International Conference on Intelligent Agent Technology (IAT 2004), pp. 295–301. IEEE Computer Society, Beijing (2004)
15. Lopez-Carmona, M.A., Velasco, J.R.: An expressive approach to fuzzy constraint based agent purchase negotiation. In: Proceedings of the International Joint Conference on Autonomous Agents and Multi-agent Systems (AAMAS 2006), Hakodate, Japan, pp. 429–431 (2006)
16. Lopez-Carmona, M.A., Velasco, J.R., Marsa-Maestre, I.: The agents' attitudes in fuzzy constraint based automated purchase negotiations. In: Burkhard, H.-D., Lindemann, G., Verbrugge, R., Varga, L.Z. (eds.) CEEMAS 2007. LNCS (LNAI), vol. 4696, pp. 246–255. Springer, Heidelberg (2007)
17. Luo, X., Jennings, N.R., Shadbolt, N., Ho-Fung-Leung, Lee, J.H.M.: A fuzzy constraint based model for bilateral, multi-issue negotiations in semi-competitive environments. Artificial Intelligence 148(1-2), 53–102 (2003)
18. Manderick, B., de Weger, M., Spiessens, P.: The genetic algorithm and the structure of the fitness landscape. In: Proceedings of the Fourth International Conference on Genetic Algorithms, San Diego, CA, pp. 1143–1150 (1991)
19. Marsa-Maestre, I., Lopez-Carmona, M.A., Velasco, J.R.: Improving trade-offs in bilateral negotiations under complete and incomplete information settings. In: Bui, T.D., Ho, T.V., Ha, Q.T. (eds.) PRIMA 2008. LNCS (LNAI), vol. 5357, pp. 275–286. Springer, Heidelberg (2008)
20. Marsa-Maestre, I., Lopez-Carmona, M.A., Velasco, J.R., de la Hoz, E.: Effective bidding and deal identification for negotiations in highly nonlinear scenarios. In: Proceedings of the VIII International Conference on Autonomous Agents and Multiagent Systems (AAMAS 2009), pp. 1057–1064 (2009)
21. Marsa-Maestre, I., Lopez-Carmona, M.A., Velasco, J.R., Ito, T., Fujita, K., Klein, M.: Balancing utility and deal probability for negotiations in highly nonlinear utility spaces. In: Proceedings of the Twenty-first International Joint Conference on Artificial Intelligence (IJCAI 2009), pp. 214–219 (2009)
22. Miller, B.L., Goldberg, D.E.: Genetic algorithms, tournament selection, and the effects of noise. Complex Systems 9(3), 193–212 (1995)
23. Nash, J.F.: Two-person cooperative games. Econometrica 21(1), 128–140 (1953)
24. Nisan, N.: Bidding languages for combinatorial auctions, ch. 9. MIT Press, Cambridge (2006)
25. Robu, V., Somefun, D.J.A., La Poutré, J.A.: Modeling complex multi-issue negotiations using utility graphs. In: AAMAS 2005: Proceedings of the fourth international joint conference on Autonomous agents and multiagent systems, pp. 280–287. ACM, New York (2005)
26. Tomassini, M., Vanneschi, L., Collard, P., Clergue, M.: A study of fitness distance correlation as a difficulty measure in genetic programming. Evol. Comput. 13(2), 213–239 (2005), `http://dx.doi.org/10.1162/1063656054088549`
27. Vassilev, V.K., Fogarty, T.C., Miller, J.F.: Smoothness, ruggedness and neutrality of fitness landscapes: from theory to application, pp. 3–44 (2003)
28. Yager, R.: Multi-agent negotiation using linguistically expressed mediation rules. Group Decision and Negotiation 16(1), 1–23 (2007)

# Acting While Negotiating in the Convoy Formation Problem

Yi Luo[1] and Ladislau Bölöni[2]

[1] School of Electrical Engineering and Computer Science,
University of Central Florida, Orlando
yiluo@mail.ucf.edu

[2] School of Electrical Engineering and Computer Science, University of Central Florida
lboloni@eecs.ucf.edu

**Summary.** In the convoy formation problem, two embodied agents are negotiating the synchronization of their movement for a portion of the path from their respective sources to destinations. We consider a setting where the negotiation happens in physical time, thus the agents have the opportunity to perform actions while negotiating. Thus, the agent's behavior is controlled by the interacting pair of negotiation and action strategies. After considering the challenges of acting while negotiating for the general convoy formation problem, we focus on a specific case where convoys can traverse a rectangular obstacle which is unaccessible to individual agents. We propose a general framework for building interacting negotiation and action strategies based on the selfishness and the optimism parameters. We propose two strategies with minimal opponent modeling, and a more complex strategy which uses particle filters to create a time evolving opponent model. Through a series of experiments we study the interaction between the negotiation and action strategies and compare the performance of the proposed strategy pairs in incomplete information scenarios.

## 1 Introduction

Let us start by defining the convoy formation problem for embodied agents. Two agents $A$ and $B$ move from their source positions $S_A$ and $S_B$ to their destinations $D_A$ and $D_B$. We assume that the agents move along the paths given by the function $P_A(t) \rightarrow L$, which we read by saying that agent $A$ is at the location $L$ at time $t$.

At the initial timepoint $t_0$ we have $P_A(t_0) = S_A$ and we define the *arrival time* of $A$ as the smallest time $t_{arr}$ for which $P_A(t_{arr}) = D_A$. For every path we define the *unit cost* $c_P(t)$, and the cost of a time segment $C(t_1, t_2) = \int_{t_1}^{t_2} c_P(t)dt$. Most of the time, we are interested in the cost of the path $C_P(t_0, t_{arr})$. In the simplest case we are only interested in the time to reach the destination. This corresponds to a unit cost $c_P(t) = 1$, and the cost of the path $C_P(t_0, t_{arr}) = t_{arr} - t_0$. Many environmental factors can be modeled by the appropriate setting of the unit costs. For instance, the unit cost might be dependent on the location $c_P(t) = f(P_A(t))$

T. Ito et al. (Eds.): Innovations in Agent-Based Complex Automated Negotiations, SCI 319, pp. 129–146.
springerlink.com 

or on the speed of the agent $c_P(t) = f(P'_A(t))$. Locations or speeds which are unfeasible to the agent can be set to have an infinite unit cost.

Two agents form a *convoy* if they are following the same path $P_{A+B}(t)$ over the period of time $[t_{join}, t_{split}]$. An agent is motivated to join a convoy because of the *convoy advantage*: the unit cost for the convoy is smaller than for the individual agent over the same path. One example is the case when convoys can traverse areas which are not accessible to individual agents: $\exists t \in [t_{join}, t_{split}]$ $\exists l$ $P_{A+B} = l$ with $c_{P,A}(t) = \infty$ and $c_{P,A+B}(t) = c \in \mathbb{R}$. Naturally, convoy and non-convoy segments of the path need to be contiguous in space: $P_A(t_{join}) = P_B(t_{join}) = P_{A+B}(t_{join}) = L_{join}$ and $P_A(t_{split}) = P_B(t_{split}) = P_{A+B}(t_{split}) = L_{split}$. We call $L_{join}$ and $t_{join}$ the join locations and time, and $L_{split}$ and $t_{split}$ the split locations and time, respectively.

We are considering self-interested agents which are searching for the path with the smallest cost from source to destination. This path might or might not include segments traversed as a convoy. In the following we assume that the agents are using negotiation to agree on the segment traversed as a convoy. The negotiation succeeds if an agreement is reached over a quadruplet $(L_{join}, t_{join}, L_{split}, t_{split})$. Convoy negotiation is thus a multi-issue negotiation, with two temporal and two spatial issues. It can be seen as a six-issue negotiation if we consider the spatial location $L = (x, y)$ as two issues.

In [6, 7] we have considered a simplified convoy formation problem called Children in the Rectangular Forest (CRF), where the convoy advantage is represented by the convoy's ability to traverse a rectangular obstacle which is not accessible to the individual agents. The CRF problem presents many challenges of the general problem such as the difficulty of establishing whether an offer is feasible to the opponent, whether it represents a concession or not, and the difficulty of simultaneously negotiating temporal and spatial issues. At the same time, the CRF problem simplifies away the path planning problem, as all the Pareto-optimal deals correspond to paths formed of at most three linear segments.

The work described in this paper represents a step towards bringing convoy negotiation closer to a more realistic setting. Rather than assuming that the agents are negotiating instantaneously, we assume that the negotiation process is happening in physical time, during which the agents can take real world actions, such as moving towards their destination, their expected meeting point or other locations. The immediate consequence is that in addition to the negotiation strategy, the agents also need to consider the *action strategy*. The relationship between the two is complex. A good action strategy will consider the current status of negotiation; in its turn, the actions taken by the agent will change the value of the exchanged offers.

The remainder of this paper is organized as follows. Section 2 discusses related work. Section 3 introduces some general considerations about the acting while negotiating for the convoy formation problem. In Section 4 we define the acting while negotiating (AWN) in the context of the CRF problem. Section 5 describes the selfishness/optimism meta-strategy for the AWN problem and two implementations with minimal opponent modeling. Section 6 describes a negotiation strategy which uses a particle filter to trace the dynamically evolving opponent model. In Section 7 we describe several experiments studying the interaction between the action and negotiation strategies, as well as comparing the different action and negotiation strategy pairs. We conclude in Section 6.

## 2 Related Work

There is a wide literature on multi-issue negotiations under incomplete information. Negotiating in the spatio-temporal domain, however, has received relatively little attention. In the following, we briefly review some of the recent papers which overlap in some aspects with our work, and whose approach (even when applied in a different domain), influenced our approach.

In the value oriented domain, the most frequently taken approach for modeling the overlap between the negotiation time and physical time is the *deadline model*. In these cases, agents do not need a separate action strategy, but they need to take the time in consideration in their negotiation strategy.

Sandholm and Vulkan [9] analyze the problem of negotiating with internal deadlines where the deadlines are private information of the agents. The negotation problem is a "split a single pie", zero-sum negotiation. They find that for rational agents, the sequential equilibrium strategy requires the agents to make no concessions until their deadline; then, the agent with the earliest deadline concedes the whole issue under negotiation.

Fatima, Wooldridge and Jennings [2] study three different negotiation procedures for multi-issue negotiation under deadlines. In the package deal procedure, all issues are discussed together, in the simultaneous procedure, issues are discussed independently but simultaneously, while in the sequential procedure, issues are discussed one after another. The authors find that the package deal is the optimal procedure for both agents.

Another aspect of the problem considered by us is the incomplete information regarding the utility and feasibility of the offers to the opponent. Information incompleteness can span a very wide range of aspects. In a simple value oriented negotiation, one might not know the opponents utility function or strategy, but usually all offers are feasible, and very often, there is no uncertainty over whether an offer is a concession or not. Information uncertainty is an additional problem when the utility function is non-linear.

Golfarelli et al. [3, 4] consider robotic agents which are assigned tasks associated with physical locations. The tasks carry precedence constraints (execute one specific task earlier than the other) and object constraints (fetch the object in order to execute the task). Agents can collaborate by swapping tasks through an announcement-bid-award mechanism to reduce the execution cost of the tasks.

Saha and Sen [8] consider the problem of negotiating efficient outcomes in a multi-issue negotiation where some of the parameters of the agent are not common knowledge and distinguish between "distributive" and "integrative" scenarios.

Crawford and Veloso [1] consider the multi-agent scheduling problem, where the agent can dynamically learn the opponents behavior and choose between a strategies proposed by a number of distinct "experts". The performance is measured in terms of total utility achieved over a series of trials.

Ito et al. [5] consider inter-dependent multi-issue negotiations which allow very complex nonlinear utility functions. The proposed solution is a bidding based negotiation protocol, in which each agent creates bids by sampling its own utility space. The mediator is responsible to identify a final contract by finding all the combinations of bids that are mutually consistent.

Tykhonov and Hindriks [10] use Bayesian learning to study the opponent's preference for a specific issue. They form a series of discrete hypotheses about the type of the opponent, with associated probabilities. These probabilities are updated based on Bayes' rule and the distance between the expected utility of the opponent's bid and the utility of actual bid.

# 3 General Considerations

## 3.1 Possible Action Strategies

The idea that a negotiation is a process which is happening in physical time is not new, but in many applications it is considered under strong simplifying assumptions. For instance, the split the pie game, used to model worth oriented negotiations, frequently assumes that the pie shrinks a fixed fraction at every negotiation round. Although this is a good model for motivating the agents to reach a deal as soon as possible, it does not capture the ability of the agents to take actions, and the relationship between the elapsed time and the value of offers is unrealistically simple.

In the case of the convoy formation problem, allowing acting while negotiating means that we consider every negotiation turn to take a time $t_r$, during which the agents can move on any feasible trajectory. For the remainder of this paper, we will make the assumption that $t_r$ is a constant value. Agents participating in a negotiation under these conditions need to have both a negotiation and an action strategy.

Let us now consider several extreme examples of action strategies. The simplest action strategy would be for the agent to stand still during the negotiation. The disadvantage of such approach is that the value of all possible deals will become lower with the amount of time wasted during negotiating. For instance, an agent which spends 100 seconds negotiating, finding out that no deal is possible, then moving on the conflict deal trajectory, would arrive 100 seconds later than an agent which did not even negotiate. This scenario is very similar to the "shrinking pie" scenarios in worth-oriented negotiations, which also assume that the agents do not act while negotiating.

The second strategy would be to continue on the originally established trajectory of the conflict deal. This corresponds to a *pessimistic* agent, which up to the moment when a deal is agreed upon will assume that no deal is possible. The advantage of this choice is that the agent has a guarantee that it will not fare worse than the conflict deal. Unfortunately, moving on the conflict deal trajectory will reduce the value of every offer, and it can make some offers unfeasible in the sense that the agent can not reach the proposed join location $L_{join}$ in time $t_{join}$.

At the other extreme, the agent might act *optimistically*: it can move on the shortest trajectory to the location of its own latest offer. Provided that the offer is accepted, this is the action which would provide the agent with the lowest possible cost. On the other hand, it requires a risky commitment from the agent: if no deal will be reached, or if the deal reached will be relatively far from the predicted one, the cost to destination will be actually higher than if the agent has not participated at all in the negotiation.

Between full optimism and pessimism, agents might choose a *hedging strategy* which moves at an intermediate trajectory between the conflict deal and the agent's own offer.

### 3.2 Baseline and Pragmatic Rationality

Allowing agents to act while negotiating requires us to refine our definition of rationality of a deal. At the beginning of the negotiation, at time $t_0$, the agent has a conflict deal path with cost $C_{conflict}$. According to the *baseline rationality* definition, any offer which has a higher cost than $C_{conflict}$ is not rational and it will not be accepted by the agent. If the agent is taking risks by acting optimistically, at some point in time $t_x$ it might find itself in the position that it has already incurred costs $C_x$, and the best path from current location $L_x$ to the destination will have a cost $C^x_{conflict}$. If at this moment an offer with cost $C_{offer}$ is received, it will be called *pragmatically rational* if $C_{offer} + C_x < C^x_{conflict}$ and *baseline rational* if $C_{offer} + C_x < C_{conflict}$. A rational agent will need to act based on the pragmatic rationality, as the original conflict deal alternative is not available any more at this moment in time. Occasionally, the agent might find it necessary to accept deals which are not baseline rational.

However, when we are measuring the overall performance of the negotiation strategy / action strategy pairs, the term of comparison should be the original conflict deal. In order for a strategy pair to be acceptable, it needs to be baseline rational at least in the statistical average.

## 4 Acting While Negotiating in the CRF Model

In the following we shall study the issue of acting while negotiating in the Children in the Rectangular Forest (CRF) problem, an instance of the convoy formation problem where the "convoy advantage" is the ability of the convoy to traverse a rectangular region inaccessible to the individual agents. We assume the negotiation protocol to be Simple Exchange of Binding Offers (no argumentation). We also assume a zero-knowledge environment; the only source of information of the agents is through the offers of the opponent. We will consider the cost of a path to be the time to destination along that path.

When an agent receives an offer from its negotiation partner, it first checks it for feasibility. An offer is not feasible if the agent can not reach the designated locations on time, and we will consider these offers to have a cost of $+\infty$. For an offer $\mathbf{O} = (L_{join}, t_{join}, L_{split}, t_{split})$ made at time $t_{crt}$, the agent $A$ with speed of $v_A$, source location at $L^A_{src}$, current location at $L^A_{crt}$ and destination at $L^A_{dest}$ will calculate the cost of the offer as:

$$C^A(\mathbf{O}) = \begin{cases} +\infty \text{ if } t_{crt} + \frac{dist(L^A_{crt}, L_{join})}{v_A} > t_{join} \\ +\infty \text{ if } \frac{dist(L_{join}, L_{split})}{v_A} > t_{split} - t_{join} \\ t_{split} + \frac{dist(L_{split}, L^A_{dest})}{v_A} \text{ otherwise} \end{cases} \tag{1}$$

Similarly we define the cost of the conflict deal as the time spent in the negotiation until the current moment $t_{crt}$, plus the time necessary to reach the destination from the current location $L_{crt}$ by going around the forest. Note that the cost of both the collaboration and the conflict deal depend on the state (the current time and location of the agent). As we discussed in the general convoy formation case, the pragmatic rationality of the offer is also state dependent. An offer might be pragmatically rational for an agent at a certain moment in the negotiation, even if its cost is higher than the original conflict deal cost. The opposite case is also possible: an offer which would have been favorable at the beginning of the negotiation might not be rational for the agent in the current state (for instance, if the agent is already well on its way towards the conflict deal).

At the other extreme from the conflict deal is the "ideal offer" with the cost $C^A_{best}$, which corresponds to the earliest time the agent can reach its destination, assuming an opponent which is ideally collaborative and has ideal capabilities. For a real opponent, this ideal offer might not be rational, or even feasible. We define the *utility* of an offer by the fraction of how much it can save from the cost of the conflict deal in comparison to the ideal offer.

$$U^A(\mathbf{O}) = \frac{C^A_{conflict} - C^A(\mathbf{O})}{C^A_{conflict} - C^A_{best}} \tag{2}$$

With this definition, the utility of non-rational offers is negative and the utility of non-feasible offers is minus infinity.

# 5 Strategies for the AWN Problem

## 5.1 The Selfishness-Optimism Meta-strategy

We have seen that an AWN agent requires a pair of interacting strategies for negotiating and acting. To capture the relationship between the two into an easy-to-understand framework, we propose a technique which integrates the offer acceptance decision and the action strategy into a single meta-strategy. This *selfishness-optimism meta-strategy* does not define the offer formation mechanism; this needs to be provided separately, and is normally inherited from non-AWN strategies.

The *selfishness* $\lambda$ is the lowest utility of the offer, as defined by Equation 2, which the agent is ready to accept. A fully selfish agent ($\lambda = 1$) will only accept its ideal offer, a fully benevolent agent ($\lambda = 0$) will accept any rational offer.

The *optimism* $\gamma$ governs the agent's movement and represents the amount of hedging between moving towards its own latest offer versus the conflict deal location. A fully pessimistic agent ($\gamma = 0$) assumes that there will be no deal and move on the conflict deal trajectory.

The reader might notice that this meta strategy can be immediately generalized by making the $\lambda$ and $\gamma$ parameters variable over the course of the negotiation. An agent, seeing that the opponent conceded too readily, might decide to drive a hard bargain by increasing its selfishness. An agent might make its optimism dependent on an external machine learning system which predicts the likelihood of a deal. A particularly Machiavellian agent might even make offers only to confuse the opponent and move to a predicted deal location which is far from its current offer.

For the remainder of this paper, we will assume agents with the $\lambda$ and $\gamma$ parameters fixed and determined at the beginning of the negotiation. The algorithm for selfishness-optimism meta-strategy is shown below, where $B(t)$ is the belief at time $t$, and $NEXTOFFER(.)$ is the offer formation function to calculate the next counter-offer.

**Algorithm 1.** The selfishness-optimism meta-strategy

receive($O^B_{t-1}$);
$B(t) \leftarrow UPDATE(B(t-1), O^B_{t-1})$;
**if** *isFeasible* $(O^B_{t-1})$ and $U\ (O^B_{t-1}) \geq \lambda$ **then**
    send$(O^B_{t-1})$; // form agreement
**else**
    $O^A_t = NEXTOFFER(B(t), \lambda)$;
    **if** $O^A_t == null$ **then**
        send($\emptyset$); // conflict deal
    **else**
        send($O^t_A$);
    **end if**
**end if**
move(L(t), B(t), $\gamma$);

## 5.2 Minimal Opponent Models

We assume a zero-knowledge environment: the only information the agents have about each other is extracted from the offers. Under these conditions, one of the main challenges of any convoy negotiation is offer formation. With four issues in every offer, the negotiation space is very large. There is no natural ordering of the offers, thus an agent can never be sure whether the offer it made is a concession to the opponent or not, and whether it is feasible or rational for the opponent. Finally, as the opponent is also acting while negotiating, the utility, rationality and feasibility of the offers changes with the passing of time and the actions of the opponent. All this makes it important for the agent to build a model of the opponent as part of its negotiation strategy.

In the following we consider some initial information which can be extracted from the initial offers of the agents. An offer does not immediately identify the agent's source and destination, even if the agent offers its own ideal trajectory. The factor which is relatively easy to identify is the speed capability of the agent. As every offer is binding, the first offer made by an agent will identify a minimum value on the agent's speed capability based on the speed on the common trajectory portion. Unless the agent is engaged in deceptive practices, this first offer will be based on its maximum possible speed.

The agent making the second offer can find itself in one of two possible situations. It can find that the opponent's speed is larger than its own. Then it needs to structure its counter-offer based on its own, lower speed. On the other hand, if it finds the opponent's speed to be smaller than its own capability, it will make an offer assuming the opponent's speed for the common part of the trajectory, without disclosing its own higher capabilities. Either way, by the end of the first offer exchange, the agents will know their maximum common speed, and will use this in all subsequent offers. Thus, the remainder of the offers will always be feasible for the common portion of the trajectory. It is, however, much harder to determine the current location of the opponent agent.

### 5.3 Two Offer Formation Strategies

Let us now describe two offer formation strategies which use only minimal user models.

**Monotonic Concession in Space (MCS)** calculates the next offer by conceding in terms of the location fields, towards the opponent's last offer. It is parameterized by the conceding pace at each side of the forest $(C_m, C_s)$. If the utility of the next conceding offer is below the selfishness, or no concession is possible (e.g. the opponent's last offer and the agent's last offer are identical in location), the negotiation stops with no agreement.

The MCS strategy resembles the monotonic concession strategy from single-issue worth-oriented domains. There are, however, some important differences. Conceding in the meeting and splitting location does not necessarily represent a concession in terms of utility. As the MCS agent does not know the opponent's location, the offer might not be feasible in terms of the join time.

One of the main problems of the MCS strategy is that by uniformly conceding in the two spatial components, it excludes a large parts of the negotiation space.

**Uniform Concession (UC)** strives to perform a better exploration of the offer space, while trying to present the offers in an order which would try to achieve the best deal which the opponent would still accept. UC first generates a pool of all possible offers, described as combinations of meeting and splitting location with a certain resolution, as well as possible time buffers for the meeting time. The splitting time is calculated based on the maximum common speed. Only the offers which are rational, feasible and have an utility higher than the selfishness $\lambda$ are included in the pool.

UC defines a conceding rate $\alpha$ and a current utility range (with the span of $\alpha$) for each round. When calculating the next offer, the agent only searches the offers in the current utility range for the one most similar to the opponent's offer. The utility range starts at 1 and decreases with $\alpha$ each round until the selfishness level is reached (see Algorithm 2). Thus every offer made will be a concession of about $\alpha$, in terms of the offering agent's utility. At every round, the UC selects from the current utility range the offer which is the most similar to the opponent's last offer. The similarity between two offers is defined by the sum of squared difference of each issue (see Equation 3). If the offer pool is empty, the negotiation is halted.

$$\mathbf{O_t^A} = \arg\min_{\mathbf{O}}(||\mathbf{O} - \mathbf{O_{t-1}^B}||^2) \tag{3}$$

**Algorithm 2.** Offer formation in the Uniform Concession agent

Create $Set\langle offer\rangle$ to hold all possible offers;<br>
**while** $Set\langle offer\rangle$ is empty **do**<br>
    $lower = lower - \alpha$;<br>
    **if** $lower \leq \lambda$ **then**<br>
        **return** $O_{next} \leftarrow null$;<br>
    **end if**<br>
    find all $Offer$ that $U(Offer) \in (lower, lower + \alpha)$;<br>
    add all $Offer$ in $Set\langle offer\rangle$;<br>
**end while**<br>
find most similar $Offer$ to $O_{t-1}^B$ in $Set\langle offer\rangle$;<br>
**return** $O_{next} \leftarrow Offer$;

# 6 Opponent Modeling with Particle Filters

As the opponent is moving while negotiating, in our case, opponent modeling requires not only learning the initial parameters, but also maintaining a dynamically evolving model of the opponent, a problem of *probabilistic reasoning over time*. In this section we describe the PF strategy which uses a Sampling-Importance-Resampling (SIR) particle filter to update its beliefs about the opponent, then uses a K-Means clustering technique to extract a likely hypothesis on which the offer formation is based.

## 6.1 The Particle Model

The PF strategy represents its knowledge about the opponent as a cloud of weighted particles. In the following we discuss (1) the particle representation, (2) the prediction model, describing how the particles evolve in time and (3) the sensor model, which describes how observations (which in our case are offers made by the opponent) affect the weight of the particle.

### The Particle Representation

A particle should contain all the information the learning agent needs to know about the opponent. We represent the particle $\mathbf{X_t}$ at time $t$ as a vector of its opponent's current state:

$$\mathbf{X_t} = \langle L_{src}, L_{crt}, L_{dest}, S_{id} \rangle$$

where $L_{src}$ is the source location, $L_{crt}$ is the current location, $L_{dest}$ is the destination, and $S_{id}$ is an identifier of the strategy used by the opponent. The strategy is chosen from a set of discrete strategies.

### The Prediction Model

At every negotiation round, the particle $\mathbf{X_t}$ is updated from its previous state $\mathbf{X_{t-1}}$ using the following equations:

$$\mathbf{X_t} = \begin{cases} L_{src}(t) = L_{src}(t-1) + \xi_{src} \\ L_{dest}(t) = L_{dest}(t-1) + \xi_{dest} \\ L_{crt}(t) = f(S_{id}, L_{crt}(t-1)) + \xi_{current} \\ S_{id}(t) = S_{id}(t-1) \end{cases}$$

where $f(.)$ is a function to calculate the next location according to the opponent's action strategy $S_{id}$ and its former location $L_{crt}(t-1)$ and $\xi$ is random noise generated from the two-dimensional normal distribution accounting for the uncertainty of the estimation.

### The Sensor Model

The particle weights are updated with every new observation. For each particle $i$, the PF agent calculates the probability $Pr(\mathbf{O_t}|\mathbf{X_t^i})$ that a hypothetical opponent described by the particle would make the specified offer. To do this, we first calculate the offer which would have been made by the agent described by the particle $O_{exp}(X_t^i)$ and then calculate the probability based on the difference of the real offer from the expected offer:

$$\begin{aligned} Pr(\mathbf{O_t}|\mathbf{X_t^i}) &= Pr(O_t|O_{exp}(X_t^i)) \\ &= g_4(y_j, t_j, y_s, t_s|y_j^{exp}, t_j^{exp}, y_s^{exp}, t_s^{exp}) \\ &= g(y_j|y_j^{exp})g(t_j|t_j^{exp})g(y_s|y_s^{exp})g(t_s|t_s^{exp}) \end{aligned}$$

In the formula, $(y_j, t_j, y_s, t_s)$ are the actual values in the opponent's last offer $\mathbf{O_t}$. $g_4(.)$ is the four-dimensional Gaussian p.d.f which centers at expected offer $O_{exp}(X_t^i)$. The weight of each particle is updated accordingly:

$$w_i(t) = Pr(\mathbf{O_t}|\mathbf{X_t^i})w_i(t-1)$$

The particle weights are normalized after the update, and if the estimate of effective number of particles

$$\hat{N}_{\mathit{eff}} = \frac{1}{\sum_{i=1}^{P}(w_i)^2}$$

is less than the threshold $N_{threshold}$, we resample using the stratified resampling algorithm.

### 6.2 Offer Formation

Algorithm 3 describes the calculation of the next offer by the PF agent. First, we associate an offer to every particle. By making the assumption that the particle is correct, we generate the offer the same way as if we would have a full-knowledge negotiation: the offer will be feasible to both agents and have a utility larger than their respective selfishness levels. If more than one such offer can be generated, we choose the one which provides largest utility to the opponent.

**Algorithm 3.** Offer formation in the PF agent

**for all** particles $i$ **do**<br>
    search all $O^i$ where $U_{agent}(O^i) \geq \lambda$ and $U_{opponent}(O^i) \geq \lambda_i$;<br>
    **if** not found $O^i$ **then**<br>
        $O^i_{best} \leftarrow null$;<br>
    **else**<br>
        $O^i_{best} \leftarrow \arg\max_{O} U_{opponent}(O^i)$;<br>
    **end if**<br>
**end for**<br>
**if** no particle has $O^i_{best}$ or $\sum w_i \leq threshold$ **then**<br>
    return $O_{next} \leftarrow null$;<br>
**else**<br>
    cluster all particles whose $O_{best} \neq null$;<br>
    calculate weights of all clusters;<br>
    find the cluster $j$ with the highest weight;<br>
    return $O_{next} \leftarrow O_{ave}(j)$;<br>
**end if**

If none of the particles has a feasible associated offer, the PF agent breaks the negotiation. Otherwise the agent proceeds to choose an offer based on the offers associated with the particles. Calculating the mean across all the particles is not a good choice, as the particles might represent disjoint hypotheses. By taking the average over the complete set of particles, the resulting estimate might fall in the low probability zone between hypotheses.

Our approach is to perform K-Means clustering on all the particles which have assigned offers. The distance metric used is the sum of squared difference between the issues. The cluster with the highest sum of weights is selected for offer formation. The averaged offer of the selected cluster will become the next counter-offer to the opponent.

## 7 Experimental Study

### 7.1 The Influence of the Selfishness and Optimism on the Agent Trajectories

To understand the impact of the selfishness and optimism settings on the behavior of agents, we have run a series of experiments. We considered a scenario where a mutually advantageous deal is possible. The size of the map is 600×400, with the

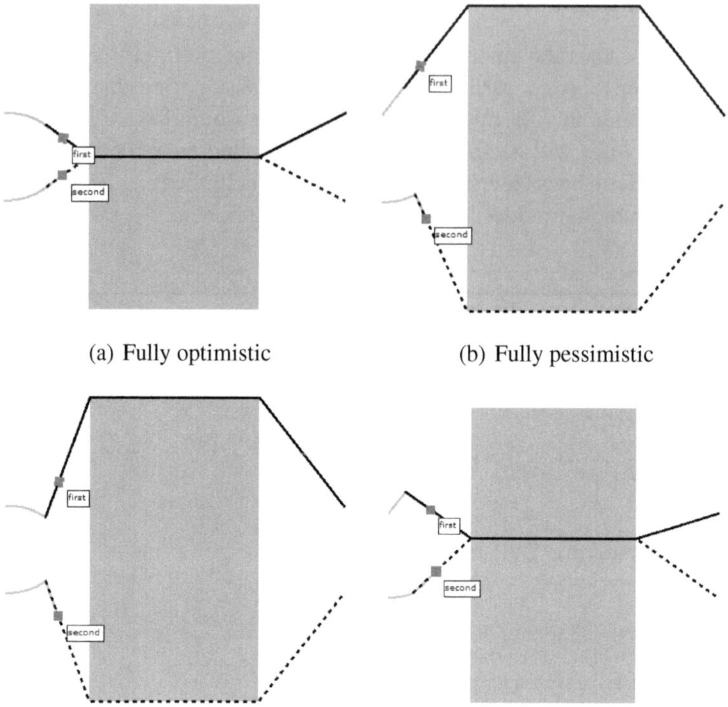


(a) Fully optimistic (b) Fully pessimistic

(c) Very selfish (d) Very generous

**Fig. 1.** The influence of the selfishness and optimism to the course and the outcome of the negotiation. The meta-strategy of agent B is fixed to $\lambda_B = 0.6$ and $\gamma_B = 1$. The values for agent A are: (a) $\lambda_A = 0.6$, $\gamma_A = 1$, (b) $\lambda_A = 0.6$, $\gamma_A = 0$, (c) $\lambda = 0.8$, $\gamma = 1$ and (d) $\lambda = 0.2$, $\gamma = 1$.

forest located at (200,25) with the size of 200 $\times$350. Agent A moves from (100,150) to (500,150) with the speed of 1.0, agent B with the fixed values of $\lambda = 0.6$ and $\gamma = 1$ moves from (100,250) to (500,250) with the speed of 1.0. Both agents use the MCS strategy($C_m = 2, C_s = 2$) to calculate the next offer. This is a "hard" scenario, because the social deal is only marginally better than the conflict deal.

Figure 1 shows the path of the agents for four different settings of the selfishness and optimism for agent A. As the MCS strategy does not depend on the current location, the actual offers exchanged are identical. Interestingly, however, in cases (a) and (d) the agents agreed to form a convoy, while for (b) and (c) they did not. Figure 1-a shows agent A with $\lambda_A = 0.6$ and $\gamma_A = 1$, that is, of average selfishness but fully optimistic. The agent moves towards its own offer at every step which results in a curving trajectory as the offer evolves. As the agents are getting closer and closer together, the utility of their respective offers keeps increasing, thus a deal is eventually reached.

In Figure 1-b agent A is fully pessimistic and of average selfishness ($\lambda_A = 0.6$, $\gamma_A = 0$). Agent A moves in a straight line towards the conflict deal, making both its own and the opponent's offers less and less valuable, despite the concessions of the opponent. Finally, the offer which the agent needs to make according to its strategy becomes of lower utility than its selfishness, the negotiation is terminated, and the opponents move on the conflict deal trajectory. Note that agent B actually ended up on a trajectory which is worse than the original conflict deal.

Figure 1-c shows a run with A being fully optimistic but of high selfishness ($\lambda_A = 0.8$, $\gamma_A = 1$). The trajectories are initially similar to case (a), however, through a series of concessions, agent A will reach a point where its next offer will have an utility smaller than its selfishness. At this point A breaks of the negotiation and moves to the conflict deal. In this case *both* agents end up on trajectories which are worse than the original conflict deal.

Finally, Figure 1-d shows a case when A is fully pessimistic but of low selfishness ($\lambda_A = 0.2$, $\gamma_A = 0$). Despite the fact that it starts to move towards the direction of the conflict deal, A and B successfully form a deal because A will accept a relatively low utility rational offer. Thus A will reverse its course and move towards the collaborative deal. Note that A had lost some utility by making the "detour" towards the conflict deal.

### 7.2 The Influence of the Action Strategy on the Offer Pool

Let us consider a negotiation turn where agent A needs to make an offer. We call the agent A's *offer pool*, the set of offers which are rational and feasible for A. The *supervisor's pool* is the set of offers which are feasible and rational for both A and B. Some strategies, such as UC generate the agent's pool explicitly. The supervisor's pool can not be computed by the agents in partial knowledge negotiations.

In the acting while negotiating problem, both the agent pool and the supervisor's pool decreases at every negotiation round, as some offers become unfeasible, as a result of the passage of time and the action strategy of the agent.

One way to characterize the agent and the supervisor pools is to consider the histogram of the offers in function of their utility. Figure 2 plots the evolution of these histograms over the negotiation scenario described in the previous section. Series of gray lines show the agent's offer pool, and black lines the supervisor's pool. Figure 2-a considers a fully pessimistic agent. As expected, the agent offer pool shrinks at every iteration. Furthermore, the maximum utility of the agent's offer pool also becomes lower at every iteration, reflecting the fact that by moving on the conflict deal trajectory, the agent is reducing its own choices. The supervisor's pool is shrinking on its own as well, and eventually becomes empty.

Figure 2-b considers a fully optimistic agent. We note that the offer pool is still shrinking at every iteration, but the amount of decrease is smaller. Furthermore, the maximum possible utility remains very close to 1.0 during the negotiation, because the agent optimistically moves towards these high utility offers. We also notice that the rate of shrinking of the supervisor's pool is much slower than in the pessimistic

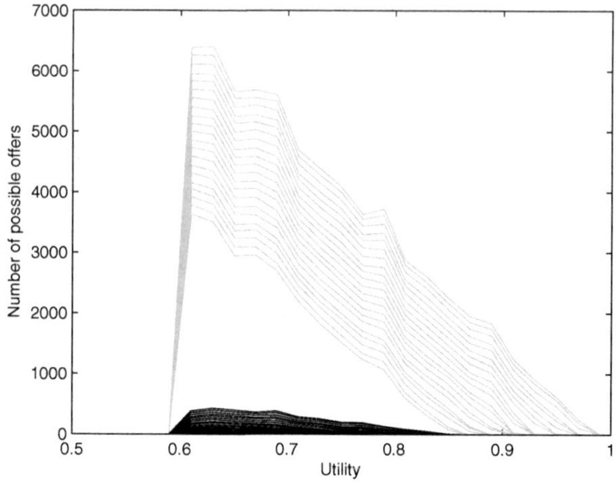


(a) Fully pessimistic

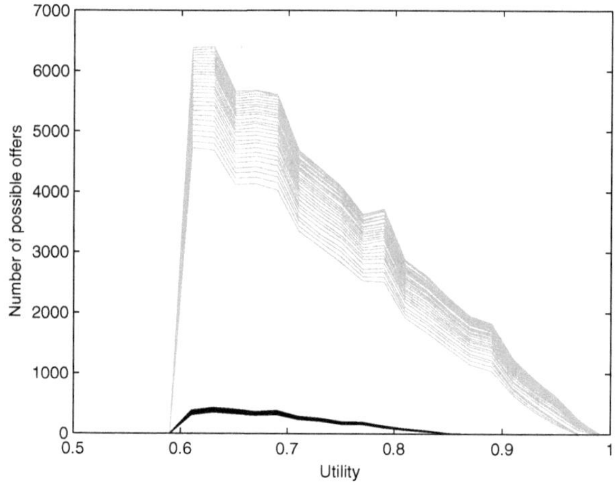


(b) Fully optimistic

**Fig. 2.** Evolution of the histograms of offer pool (gray lines) and the supervisor's pool (black lines) function of the utility. (a) $\gamma = 0$ (fully pessimistic) and (b) $\gamma = 1$ (fully optimistic). For both cases, the $\lambda = 0.6$.

case. Most of the offers which were feasible at the beginning of the negotiation remain feasible if the agent acts optimistically.

### 7.3 Opponent Modeling in the PF Agent

We will illustrate the opponent modeling in the PF agent and the overall benefits of the approach by tracing an example where the MCS agent ends in conflict, while a PF agent with the same selfishness and optimism succeeds in negotiating a deal.

Let us consider the negotiation in Figure 1-c, where A is an MCS agent with $\lambda = 0.8$ and $\gamma = 1$, which ends in conflict. We repeat the experiment, replacing the MCS agent with a PF agent, with the same $\lambda$ and $\gamma$ values.

Figure 3 shows the evolution of the particle filter through the first six steps of negotiation. In this figure, a particle is represented by two dots – one on the current

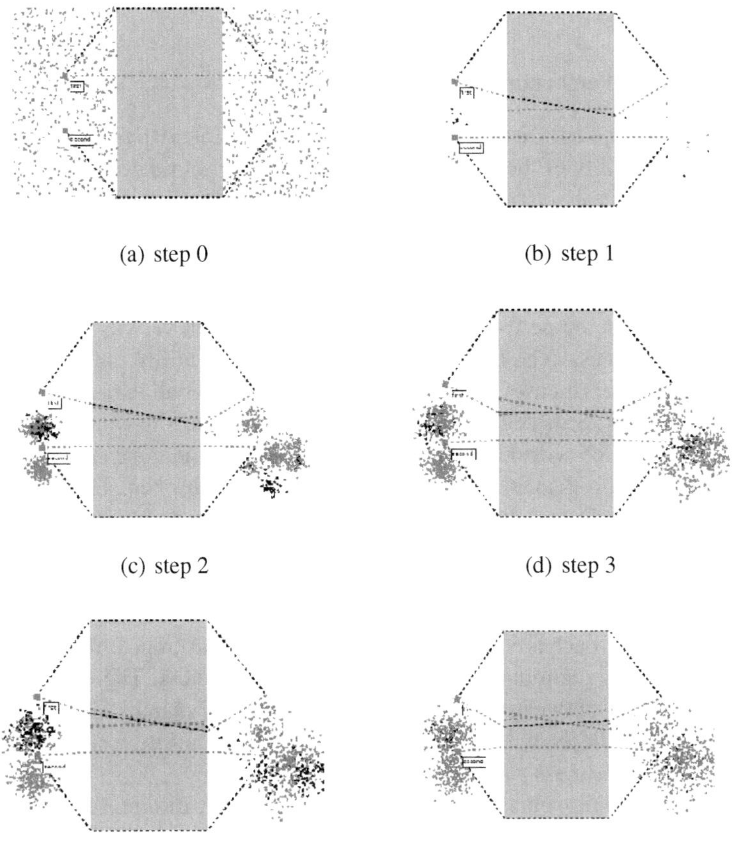

(a) step 0 (b) step 1

(c) step 2 (d) step 3

(e) step 4 (f) step 5

**Fig. 3.** Evolution of the opponent model in the PF agent. The black dots and the corresponding dashed line is the cluster selected for offer formation. The gray dots are particles belonging to the other clusters.

location side and one on the destination side. The particles from the cluster chosen for offer formation are shown in black, while the others in gray.

We note that the particles show a relatively large spread which changes from step to step. This is a result of the way in which the offers are formed based on the strongest cluster. If the opponent declines the offer, this represents a strong negative feedback to the selected cluster. This leads to a large variation in the particle cloud, further amplified by the resampling step. Nevertheless, the particle clouds track relatively well the current location and destination of the opponent, which allows the PF agent to choose better offers from the offer pool. In our case, at negotiation round 10, the opponent accepts the PF agents offer, and they move together to their meeting location. Thus, the PF agent, under the same selfishness and optimism parameters, and starting from zero knowledge, could "save" a deal, which was lost for a MCS agent using the same parameters.

### 7.4 Statistical Performance Comparison

The quality of a specific action strategy / negotiation strategy pair can be measured by the average utility of the deals it can reach over a set of randomly chosen representative scenarios against specific opponents. The statistical averaging is necessary because some strategies might be a better fit for certain scenarios: for instance, fully pessimistic action strategies will yield the best performance in scenarios where no deal is possible.

Figure 4 shows the performance of four strategies with various values for optimism and selfishness. The top figure shows the relative utility obtained while the bottom the number of cases where a deal was formed. For all experiments, the opponent uses the MCS strategy with $\lambda = 0.6$ and $\gamma = 1$, The four strategies are MCS, UC and PF to which we add MCSN, a variant of MCS where the action strategy is to not move until a deal is agreed or the negotiation is broken. Thus, the MCSN agent does not perform acting while negotiating, and the optimism parameter has no impact in this case.

The first obvious conclusion is that the all the proposed acting while negotiating strategies outperform the MCSN "don't act" strategy. There is also a clear advantage of UC and PF approaches compared to MCS in terms of average utility and number of deals, for every combination of optimism and selfishness. There is a relatively smaller difference between PF and UC. The PF strategy obtains a higher percentage of successful deals and it achieves a higher average utility for the majority of optimism and selfishness values.

Different strategies obtain their maximum utilities at different selfishness and optimism values. For MCS this value is at $\lambda = 0.8$ and $\gamma = 0$. For PF and UC it is around $\lambda = 0.6$ and $\gamma = 0.2$. Note, however, that these values are dependent on the opponent, and further studies with a range of opponents are necessary before definitive conclusions can be drawn.

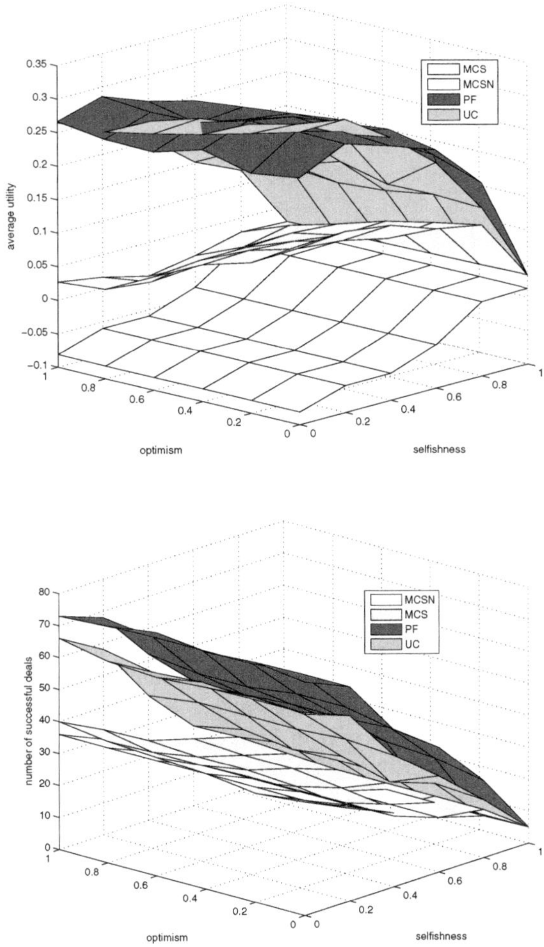


**Fig. 4.** Relative performance of various strategies negotiating with (another MCS agent) over 100 scenarios. Top: average utility, bottom: number of successful deals

## 8 Conclusions

In this paper we introduced the acting while negotiating variant of the convoy formation through negotiation problem. We have identified that the main challenge of this problem is the interaction between the negotiation strategy and the action strategy. We have introduced several negotiation and action strategy pairs both without and with opponent modeling capabilities.

Our future work involves both extending the proposed strategies to more general settings, as well as in developing more complex action strategies, such as strategies where the opponent model is used to adjust the optimism of the action strategy.

**Acknowledgements.** This work was partially funded by NSF Information and Intelligent Systems division under award 0712869.

## References

1. Crawford, E., Veloso, M.: An experts approach to strategy selection in multiagent meeting scheduling. Autonomous Agents and Multi-Agent Systems 15(1), 5–28 (2007)
2. Fatima, S.S., Wooldridge, M., Jennings, N.R.: Multi-issue negotiation with deadlines. Journal of Artificial Intelligence Research 27, 381–417 (2006)
3. Golfarelli, M., Maio, S.R.D.: Multi-agent path planning based on task-swap negotiation. In: Proc. of the 16th UK Planning and Scheduling SIG Workshop, pp. 69–82 (1997)
4. Golfarelli, M., Rizzi, S.: Spatio-temporal clustering of tasks for swap-based negotiation protocols in multi-agent systems. In: Proc. of the 6th Intl. Conf. on Intelligent Autonomous Systems, pp. 172–179 (2000)
5. Ito, T., Hattori, H., Klein, M.: Multi-issue negotiation protocol for agents: Exploring nonlinear utility spaces. In: Proc. of the Intl. Joint Conf. on Artificial Intelligence, pp. 1347–1352 (2007)
6. Luo, Y., Bölöni, L.: Children in the forest: towards a canonical problem of spatio-temporal collaboration. In: Proc. of the Sixth Intl. Joint Conf. on Autonomous Agents and Multi-Agent Systems (AAMAS 2007), pp. 986–993 (2007)
7. Luo, Y., Bölöni, L.: Collaborative and competitive scenarios in spatio-temporal negotiation with agents of bounded rationality. In: Proc. of the 1st Int. Workshop on Agent-based Complex Automated Negotiations (ACAN 2008), pp. 40–47 (2008)
8. Saha, S., Sen, S.: Negotiating efficient outcomes over multiple issues. In: Proc. of the 5th Intl. Joint Conf. on Autonomous Agents and Multiagent Systems (AAMAS 2006), pp. 423–425 (2006)
9. Sandholm, T., Vulkan, N.: Bargaining with deadlines. In: Proc. of Natl. Conf. on Artificial Intelligence (AAAI 1999), pp. 44–51 (1999)
10. Tykhonov, D., Hindriks, K.: Opponent modelling in automated multi-issue negotiation using bayesian learning. In: Proc. of the Seventh Intl. Conf. on Autonomous Agents and Multiagent Systems (AAMAS 2008), pp. 331–338 (2008)

# Multi-player Multi-issue Negotiation with Complete Information

Mengxiao Wu[1], Mathijs de Weerdt[2], Han La Poutré[1], Chetan Yadati[2], Yingqian Zhang[2], and Cees Witteveen[2]

[1] Centre for Mathematics and Computer Science (CWI), Amsterdam, The Netherlands
wu@cwi.nl, hlp@cwi.nl
[2] Delft University of Technology, Delft, The Netherlands
M.M.deWeerdt@tudelft.nl, c.yadati@tudelft.nl, Yingqian.Zhang@tudelft.nl, C.Witteveen@tudelft.nl@tudelft.nl

**Summary.** This paper presents a multi-player multi-issue negotiation model to solve a resource allocation problem. We design a multilateral negotiation protocol, by which rational players bid sequentially in consecutive rounds till a deadline. Every player's bid is a proposal for resources allocated to himself. In this framework, we perform a thorough theoretical analysis of negotiation with complete information, which is also a preliminary for the more complex incomplete information case. Here we extend known results on single issue bilateral negotiations to multiple players and multiple issues. We show that, under a complete information setting, we can derive the negotiation strategies which form a subgame perfect equilibrium. We also show that when a discount factor exists, an agreement will be reached immediately at the end of the first negotiation round. The outcome is Pareto optimal, when every player maximizes his utility by making trade-offs between issues and selects the bid that is best for his opponents amongst multiple bids that would give the same maximum utility to himself.

## 1 Introduction

With the rapid development of multi-agent systems, automated negotiation has been widely used to solve coordination and cooperation problems in complex systems. In this paper, we propose a solution when multiple players allocate multiple resources amongst themselves through negotiation. In contrast to most previous work on two-player multi-issue negotiation [6] or multi-player single-issue negotiation [2, 9, 16], the negotiation model presented in this work is a *multi-player multi-issue* strategic negotiation model. It is also different from the model of multiple bilateral negotiation between more than two players [8]; it is a *multilateral* negotiation that always involves all players in a single negotiation.

We design a negotiation protocol where each player bids a desired allocation of resources only for himself. Compared to Rubinstein's alternating-offer bargaining [14], in which one player's proposal includes the allocations for all players, this model applies to many real negotiation scenarios and the equilibrium solution

T. Ito et al. (Eds.): Innovations in Agent-Based Complex Automated Negotiations, SCI 319, pp. 147–159.
springerlink.com 

applies to those scenarios directly. Fatima *et al* [6] study different approaches to multi-issue negotiation and conclude that the package approach, i.e. when all issues are bundled and negotiated concurrently, is the optimal way for multi-issue negotiation. In our work, we extend this concurrent multi-issue negotiation model from two players to multiple players, in a model where we let every bid made by a player include the desired allocation of all resources.

The model presented in this paper tackles several problems introduced by multi-player negotiation. First, we discuss the bidding order problem in the negotiation between more than two players and provide a simple way to solve it. Second, we describe the reasoning process behind the negotiation strategies for more than two players, present the resulting strategies, and give mathematical proofs of their properties. Although we just analyze the negotiation with *complete information* in this paper, the proposed model is a fundamental result of automated negotiation. This paper thus is an important step towards the incomplete information case and provides a benchmark for multi-player multi-issue negotiation.

We briefly describe the negotiation model here and present the full details in the next sections. We study solutions with $n \geq 2$ players to allocate $m \geq 2$ resources amongst themselves through negotiation, which takes place round by round under a time constraint, a negotiation deadline. All players have to reach an allocation agreement, otherwise no resource will be allocated. Different from the two-player negotiation where one player's allocation determines his opponent's allocation indirectly, in the $n$-player negotiation, even if one player's allocation is determined separately, the rest of the players still need to negotiate the allocations of the resources left. Instead of letting one player propose the allocations for all players as done in [2, 14], we let all players sequentially propose only the resources to be allocated to himself, because this is all they are interested in. We set the negotiation under a complete information environment, in which all information is common knowledge, and derive equilibrium strategies of the players which form a *subgame perfect equilibrium* (SPE) [15] from *two-dimensional* reasoning, which will be discussed later. Given another time constraint, a discount factor that decreasing the utilities of agreements during the negotiation, an agreement can be reached immediately at the end of the first round. Further, the solution is *Pareto optimal* when each player selects the bid that is best for his opponents amongst multiple bids that would give the same maximum utility to himself.

The rest of the paper is organized as follows. Section 2 describes the negotiation model including the problem model, the negotiation protocol and utility functions. Section 3 proposes the equilibrium strategies formally and proves several properties of the outcome. A simple example of the equilibrium strategies is also given. Section 4 gives a brief summary of related work. And finally, Section 5 presents the conclusions and future work.

## 2 The Negotiation Model

Suppose a complex task requires a finite set of $n \geq 2$ players to perform, given a finite set of $m \geq 2$ divisible resources. We assume every player has a monotonically increasing utility for the allocated resources. Every player's allocation can only be implemented if all players agree with it. To solve the problem of resource allocation, we propose a solution where $n$ players allocate $m$ resources amongst themselves through multilateral negotiation. For the sake of simplicity, we interpret an amount $x_k$ as an allocation percentage of the $k^{th}$ resource where $x_k \in [0, 1]$ and $1 \leq k \leq m$. We denote the complete set of the total amounts of resources by a vector **1** in which every element is 1. In the rest of this paper, we use the term issue to indicate the amount of a resource negotiated by the players.

## 2.1 The Negotiation Protocol

In this section, we impose a negotiation protocol that describes how players can act and interact during the negotiation. We let the negotiation take place round by round $r \in \mathbb{N}$, in which the players can take actions. There are two common time constraints, a negotiation deadline $\gamma \in \mathbb{N}$ and a constant discount factor $\delta \in (0, 1)$. If the players cannot reach an allocation agreement on all issues in any round $r \leq \gamma$, the negotiation fails and all players get nothing. We let **disagreement** denote this outcome, which is the worst outcome of this negotiation. Given an allocation agreement at a subsequent moment, the utility that a player gets is decreased by the discount $\delta$.

We let $n$ players take actions sequentially in consecutive rounds till the deadline $\gamma$. Different from Rubinstein's alternating-offer bargaining [14], in each round $r \leq \gamma$, we let each player bid one desired combination of the $m$ issues for himself sequentially, given the bidding order of the current round $r$. In this work, the negotiation protocol requires the bidding orders of all rounds to be pre-specified and fixed during the whole negotiation, but the way to generate the bidding orders can be various. Because the order can be different in each round, we let a given player be represented by a different bidder in each of the rounds. For instance, in the case of three players 1, 2 and 3, the bidding orders can be $\langle(1, 2, 3), (2, 3, 1), (3, 1, 2), \ldots\rangle$, in which player 1 is represented by the first bidder, the third bidder and the second bidder in rounds $1, 2$ and 3 respectively. We let bidder $i \in \mathbb{N}$ represent the $i^{th}$ bidder in a round where $1 \leq i \leq n$ and let $N$ denote the set of bidders $\{1, \ldots, n\}$ of a round.

When it is a bidder's turn to bid in round $r \leq \gamma$, given the bids of the previous bidders in the current round, the bidder can either accept those bids and make his own bid, or reject those bids, in which case the negotiation passes on to the next round. At the beginning of every round $r$, the issues available for bidders to bid are always the complete set of issues **1**. The issues cannot be bidden separately. The bidder either bids a combination of all $m$ issues or rejects to bid for any issue. Therefore, the set of all possible bids is $B = [0, 1]^m$ and bidder $i$'s bid is an $m$-vector $\mathbf{x}_i \in B$. An element[1] $x_{i,k} \in \mathbf{x}_i$ represents the $k^{th}$ issue of bid $\mathbf{x}_i$ where $1 \leq k \leq m$. We let **reject** denote the action of rejection. Therefore, the set of all possible actions of every bidder is $A = B \cup \{\textbf{reject}\}$. We let $a_i \in A$ denote bidder

[1] In this paper, we also use $\in$ to represent the relation that an element belongs to a vector.

$i$'s action in a round and let $\mathbf{a} \in A^n$ denote an action profile chosen by $n$ bidders in the round. We define the constraint $\mathcal{C}$ of an **agreement** to be:

$$\textbf{reject} \notin \mathbf{a} \text{ and } \mathbf{a} = (\mathbf{x}_1, \ldots, \mathbf{x}_n) \text{ subject to } \forall k \in \{1, \ldots, m\} \sum_{i=1}^{n} x_{i,k} \leq 1$$

Given the definitions above, we propose the negotiation protocol.

- In each negotiation round $r \leq \gamma$, from the first bidder to the last bidder, every bidder $i \in N$ takes an action $a_i \in A$ sequentially.
- In round $r < \gamma$, given all previous bids $(\mathbf{x}_1, \ldots, \mathbf{x}_{i-1})$, bidder $i$ can either accept them and bid $\mathbf{x}_i$, or **reject** them. If bidder $i$ bids $\mathbf{x}_i$, then it is bidder $i+1$'s turn to choose his action/response. If bidder $i$ chooses to **reject** the previous bids, the current round $r$ ends and the negotiation passes on to round $r+1$. Once bidder $n$ accepts all previous bids and bids $\mathbf{x}_n$ and the action profile $\mathbf{a}$ satisfies the constraint $\mathcal{C}$, an **agreement** is reached and the negotiation stops successfully.
- If no agreement is reached in any round $r \leq \gamma$, the negotiation stops unsuccessfully and the outcome is **disagreement**.

## 2.2 Utility Functions

As defined in the last section, the outcome of the negotiation is some **agreement** or **disagreement**. Each player's preferences over the outcomes is represented by a utility function, which is common knowledge in this game. In the following, we define all functions for a bidder. Recall however that the mapping of a player and the bidder representing him in each negotiation round is specified by the bidding orders.

Because of the discount factor $\delta$, an agreement reached in different rounds introduces different utilities to the players. Hence, the utility depends not only on the action profile but on the round number as well. We define the utility function $u_i : A^n \times \mathbb{N} \rightarrow \mathbb{R}$, where $u_i(\mathbf{a}, r)$ represents the utility that bidder $i \in N$ would get in round $r \in \mathbb{N}$, if the bidders all chose their actions as specified in $\mathbf{a} \in A^n$. Note that **disagreement** only happens at the end of the negotiation deadline $\gamma$, when the action profile $\mathbf{a}$ does not satisfy the constraint $\mathcal{C}$, and its utility is normalized to zero. An **agreement** may be reached in any round $r \leq \gamma$, when the action profile $\mathbf{a}$ satisfies the constraint $\mathcal{C}$. In this situation, the utility that bidder $i$ would get through action profile $\mathbf{a}$ only depends on his own bid $\mathbf{x}_i \in \mathbf{a}$; we define a general valuation function $v_i : B \rightarrow \mathbb{R}$ to express this value. In this work, we assume the valuation function to be a *monotonically increasing* function of any element $x_{i,k} \in \mathbf{x}_i$ $(1 \leq k \leq m)$. Formally, the utility function is given by:

$$u_i(\mathbf{a}, r) = \begin{cases} 0 & \text{if } r = \gamma \text{ and not } \mathcal{C} \\ \delta^{r-1} \cdot v_i(\mathbf{x}_i) & \text{if } r \leq \gamma \text{ and } \mathcal{C} \end{cases} \tag{1}$$

Note that the range of $v_i(\mathbf{x}_i)$ is $\mathbb{R}$ but not $\mathbb{R}^+$, which means the value of an allocation for bidder $i$ can be negative, so the utility of an action profile $\mathbf{a}$ for bidder $i$ can be negative, if action profile **a** actually happens. However, because the rational bidders prefer **disagreement** to any allocation with a negative utility, that action profile **a** cannot really form an outcome of this game and **disagreement** is the worst outcome of this game. This makes the model more expressive than the typical problem of a cake partition. To reach an agreement, in which his opponents accept any allocation with zero utility at the end of the game, the bidding player still has to leave a minimum amount of resources to them but not nothing. This point makes the problem setting in this work relevant to more scenarios in the real world.

Before we derive the negotiation strategies, we make the following assumptions about the players and negotiation.

- *Unanimity*: only a unanimous agreement can be accepted and then be implemented.
- *Rationality*: every player will act in order to maximize his own utility.
- *Patience*: all players are patient enough to stay in the negotiation till the deadline $\gamma$, if no agreement has been reached yet.
- *Benevolence*: when a player can choose between multiple outcomes which are indifferent to him but not to his opponents, he will choose the one that is best for his opponents as far as he knows.

In this work, we assume the negotiation takes place under a complete information setting. The time constraints, the above assumptions and the preferences of players are all common knowledge.

# 3 The Negotiation Strategies

In this section, we investigate the equilibrium strategies of the players of the negotiation and use the notion of subgame perfect equilibrium [15], which induces a Nash equilibrium in every subgame (round), to examine the solution formed by those equilibrium strategies.

## 3.1 Description of Strategies

We analyze the equilibrium strategies to specify the optimal action of every bidder $i \in N$ in any round $r \leq \gamma$, when it is his turn to bid, given the previous bids in round $r$. Bidder $i$'s optimal action is to maximize the utility that he would get when the game ends. Bidder $i$'s equilibrium strategy is computed by considering all possible actions in $A$ to find the one that has the maximum utility. As specified by the utility function (1), any bid's utility for bidder $i$ is determined not only by bidder $i$'s valuation but also by whether his opponents accept it. Every bidder's optimal action is the one that maximizes his own utility with the consideration of his opponents' responses. Therefore, all bidders' optimal actions in a round are best responses to each other; the action profile forms a *Nash equilibrium* [12]. When it is bidder $i$'s turn to bid, we let $h_i = (\mathbf{x}_1, \ldots, \mathbf{x}_{i-1})$ denote the history of previous bids in the current round and let $H$ denote the set of all possible profiles of bids in the game. We define the optimal action function $s_i : H \times \mathbb{N} \rightarrow A$, where $s_i(h_i, r)$ is bidder $i$'s optimal action in round $r$, given previous bids $h_i$ in round $r$. We let $a_i^*$ denote the optimal action of bidder $i$ in the current round $r$.

We let $-i$ denote the set of all bidders other then $i$ in a round, so a combination of their actions can be represented by $a_{-i} = \times_{j \in N-i} a_j$. Given an action $a_i \in A$, bidder $i$ reasons his opponents' responses $a_{-i}$ first, and then calculates the utility of the action profile $\mathbf{a} = (a_i, a_{-i})$ for himself. The utility function (1) only gives the utilities of an action profile $\mathbf{a} \in A^n$ for every player, when $\mathbf{a}$ forms either the outcome of an **agreement** or **disagreement** in round $\gamma$. However, when action profile $\mathbf{a}$ contains a **reject** in the current round $r < \gamma$ and the negotiation passes on to the next round $r+1$, the utilities that the players would get are not specified. We think the utility of an action profile $\mathbf{a}$ for bidder $i \in N$ in round $r \leq \gamma$ to be equal to the utility that the player represented by bidder $i$ in round $r$ would get in the next round $r+1$, if action profile $\mathbf{a}$ does not form an **agreement** in the current round $r$. We define a function $w_i : A \times H \times \mathbb{N} \to \mathbb{R}$, where $w_i(a_i, h_i, r)$ represents the utility that bidder $i$ would get through action $a_i$ in round $r$, given the history $h_i$.

When it is bidder $i$'s turn to bid in round $r \leq \gamma$, bidder $i$ needs to reason the utility that he would get in round $r+1$. The result will be compared to the utility of any possible bid for him to determine his optimal action. To calculate the utility of any bid $\mathbf{x}_i \in B$, bidder $i$ needs to reason the best response of each of the remaining bidders $j > i$ in round $r$ to his own bid $\mathbf{x}_i$. The best response is bidder $j$'s optimal action derived from (i) the previous bids $(\mathbf{x}_1, \ldots, \mathbf{x}_{i-1})$, (ii) bidder $i$'s possible bid $\mathbf{x}_i$, and (iii) the reasoned optimal action of bidder $j'$ where $i < j' < j$. The reasoning also requires the information of the utilities that all players would get in the next round $r+1$. Eventually, bidder $i$ in round $r$ does two-dimensional reasoning from the current round to the last round and from the first bidder to the last bidder in each round, which is a *recursive* procedure with a base case that all players will get zero utilities after round $\gamma$, if no agreement has been reached.

### 3.2 Formal Definition of Strategies

Given the description above, we formally define the optimal function and present the negotiation strategies. We develop some notations first. As the presentation of the strategies is concerned with the consecutive rounds, we let $r$ and $r+1$ denote the current round and the next round respectively. We use a letter and the letter with a tilde to denote a bidder of round $r$ and a bidder of round $r+1$ respectively, which represent the same original player. For instance, bidders $i$ and $\tilde{\imath}$ denote the $i^{th}$ and $\tilde{\imath}^{th}$ bidders in rounds $r$ and $r+1$ respectively; they have the same utility function as they represent the same player.

Formally, given the history of previous bids $h_i = (\mathbf{x}_1, \ldots, \mathbf{x}_{i-1})$, the optimal action function is defined by:

$$s_i(h_i, r) \in \operatorname*{argmax}_{a_i \in A} w_i(a_i, h_i, r)$$

where

$$w_i(a_i, h_i, r) = \begin{cases} 0 & \text{if } r > \gamma \\ u_i(\mathbf{a}, r) & \text{if } r \leq \gamma \text{ and } \mathcal{C} \\ w_{\tilde{i}}(a_{\tilde{i}}^*, h_{\tilde{i}}, r+1) & \text{otherwise} \end{cases}$$

where

$$\mathbf{a} = (h_i, a_i, a_{i+1}^*, \ldots, a_n^*)$$
$$a_{i+1}^* = s_{i+1}(h_{i+1}, r),\ h_{i+1} = (h_i, a_i)$$
$$\forall j \in \{i+2, \ldots, n\}\ \{a_j^* = s_j(h_j, r),\ h_j = (h_{j-1}, a_{j-1}^*)\}$$

$$\forall \tilde{j} \in N \left\{a_{\tilde{j}}^* = s_{\tilde{j}}(h_{\tilde{j}}, r+1), h_{\tilde{j}} = (h_{\tilde{j}-1}, a_{\tilde{j}-1}^*)\right\}. \tag{2}$$

In any state of any round $r$, when it is bidder $i$'s turn to bid, he uses the above optimal function to calculate the optimal bid/response, given the previous bids in the current round $r$.

We let $S_i^r$ denote the equilibrium strategy of a player when he is represented by bidder $i \in N$ in round $r \leq \gamma$, let $S^r$ denote the equilibrium strategies in round $r$ where $S^r = (S_1^r, \ldots, S_n^r)$, and let $S = (S^1, \ldots, S^\gamma)$ denote the strategy profile of the players of this game.

**Proposition 1.** *The equilibrium strategy of bidder $i \in N$ in round $r \leq \gamma$ is $S_i^r$, which is given by Algorithm 1. The strategy profile $S = (S^1, \ldots, S^\gamma)$ induces a subgame perfect equilibrium of the game. If an agreement exists in this game, it will be reached immediately at the end of round 1.*

---

**Algorithm 1.** $S_i^r$ $(i \in N, r \leq \gamma)$

**Input:** previous actions $h_i = (\mathbf{x}_1, \ldots, \mathbf{x}_{i-1})$
**Output:** optimal action $a_i^*$
$a_i^* = s_i(h_i, r)$
**if** $a_i^* = \mathbf{x}_i \in B$ **then**
  **bid** $\mathbf{x}_i$
**else**
  **reject** $h_i$
**end if**

---

*Proof.* We argue that the strategy profile $S$ forms a subgame perfect equilibrium, so we need to show that $S$ induces a Nash equilibrium in every subgame (round). We are going to prove that the optimal action of any bidder $i \in N$ in any round $r \leq \gamma$ given by the equilibrium strategy $S_i^r$ is the best response to the optimal actions of his opponents $-i$ given by the equilibrium strategies $S_{-i}^r$. We give a proof by contradiction.

In any round $r$, when it is bidder $i$'s turn to bid, his optimal action is $a_i^* = S_i^r$, given the previous bids induced by the equilibrium strategies. We let $a_{-i}^*$ denote action profile induced by the equilibrium strategies $S_{-i}^r$, given $a_i^*$. Suppose any other strategy used by bidder $i$ is to choose another action $a'_i \in A$ where $u_i((a'_i, a'_{-i}), r) > u_i((a^*_i, a^*_{-i}), r)$, when all players other than $i$ adhere to $S^r_{-i}$ which induces the action profile $a'_{-i}$, given $a'_i$.

When the profile of optimal actions $\mathbf{a}^* = (a^*_i, a^*_{-i})$ can form an agreement in round $r$, the utility $u_i(\mathbf{a}^*, r)$ has been maximized in Equation (2) while the utility $u_j(\mathbf{a}^*, r)$ is no less than the utility that bidder $j$ would get in round $r+1$ where $j \in N$. Because all utility functions are monotonically increasing, if the action profile $\mathbf{a}' = (a'_i, a'_{-i})$ is a profile of bids and increases the utility for bidder $i$, $\mathbf{a}'$ either violates the constraint $C$ or lets at least one of other bidders get a utility less than the utility that he would get in round $r+1$, if $\mathbf{a}'$ is really implemented. Hence, the action profile $\mathbf{a}'$ cannot form an agreement in round $r$. The utility that bidder $i$ can get by taking action $a'_i$ in round $r$ equals to the utility that he would get in round $r+1$, which is no more than the utility $u_i(\mathbf{a}^*, r)$. There is a contradiction. Because the utility function defined in this work is completely general, it is possible that no agreement exists in the game. For any action profile, every player gets the same utility, zero, so there is also a contradiction. Therefore, the optimal action $a^*_i$ is bidder $i$'s best action/response to his opponents' actions induced by the equilibrium strategies. The equilibrium strategies $S^r$ induces a Nash equilibrium in round $r$ and the strategy profile $S$ induces Nash equilibrium in every round $r \leq \gamma$, which is a subgame perfect equilibrium.

When the strategy profile $S$ can form an agreement $\mathbf{a}$ in round $r \leq \gamma$, then every bidder $i \in N$ gets a utility $u_i(\mathbf{a}, r)$, which is no less than the utility that he would get in the next round $r+1$. The action profile $\mathbf{a}$ can also be an agreement in round $r+1$. Because of the discount factor $\delta$, at least one bidder $i$ in round $r$ has $u_i(\mathbf{a}, r) > u_{\bar{i}}(\mathbf{a}, r+1)$ and any other bidder $j$ in round $r$ has $u_j(\mathbf{a}, r) \geq u_{\bar{j}}(\mathbf{a}, r+1)$. Under the assumption of benevolence, an agreement reached earlier is always preferred by all players. Thus, an agreement will be reached immediately at the end of round 1 and the negotiation stops. □

In the optimal action function (2), it is possible that bidder $i \in N$ has multiple bids that have the same maximum utility, which are indifferent to bidder $i$ but not to bidder $j > i$. There may be an opportunity to increase the outcome utility for bidder $j$ without decreasing the outcome utility for bidder $i$. Therefore, under the assumption of benevolence, if bidder $i$ has multiple bids that have the same maximum utility, we let bidder $i$ choose the one that is best for bidder $i+1$. If bidder $i$ still has more than one bid that is best for bidder $i+1$, we let bidder $i$ choose the one that is best for bidder $i+2$, etc. This selection will last until bidder $i$ has only one optimal bid left or bidder $n$ has already been considered by bidder $i$.

**Proposition 2.** *The equilibrium outcome is a Pareto optimal under the assumption of benevolence.*

*Proof.* We argue that the equilibrium outcome is Pareto optimal, so we need to prove that no other outcome can increase the outcome utility for any player without decreasing the outcome utility for any other player, when every player chooses his optimal action benevolently. We give proof by contradiction.

When the equilibrium strategies $S$ can reach an agreement $\mathbf{a}$ at the end of the first round. Suppose bidder $i \in N$ has another bid $\mathbf{x}'_i$ that the action profile $\mathbf{a}' = (\mathbf{x}'_i, \mathbf{x}'_{-i})$ also forms an agreement in that round where $u_i(\mathbf{a}', 1) > u_i(\mathbf{a}, 1)$ and $u_j(\mathbf{a}', 1) \geq u_j(\mathbf{a}, 1)$ $(j \in N - i)$. Because all utility functions are monotonically increasing and $u_i(\mathbf{a}, 1)$ has been maximized in Equation (2) when every player has already benevolently selected his bid, the bid profile $\mathbf{a}'$ either violates the constraint $\mathcal{C}$ so that $\mathbf{a}'$ is not an agreement or lets at least one bidder $j$ get $u_j(\mathbf{a}', 1) < u_j(\mathbf{a}, 1)$. There is a contradiction. Therefore, the equilibrium outcome $\mathbf{a}$ is a Pareto optimal.

When no agreement can be reached by using the equilibrium strategies $S$, every player gets zero utility. It is impossible to increase the utility of any player without rejections from other players Therefore, the equilibrium outcome introduced by $S$ is a Pareto optimal solution of the game. □

**Proposition 3.** *The negotiation mechanism is individually rational.*

*Proof.* Because whether the outcome is an **agreement** or **disagreement**, every player gets a utility no less than zero, which is also the utility for every player if he does not participate the game, the negotiation mechanism is individually rational. □

### 3.3 A Simple Example

In this section, we use a simple example to illustrate the negotiation model and equilibrium strategies. Suppose three students need to share an office. They all prefer to have the office only to themselves. They therefore decide to time-share the office, but they agree to allow the others to leave their stuff (books, etc.) behind in the cupboard. Each of the students would like to have the office as long as possible. Let therefore the first issue be the part of the working day a student has on his own. They also like to get as much space in the cupboard as possible. Let the second issue thus be the part of the cupboard they are entitled to. The dean overheard them discussing and said: "you should take turns in making proposals to each other, but if you haven't reached an agreement before noon, I'll give the room to someone else. Determine the order of the turns by drawing straws."

Given the above case, we let three players, $1, 2$ and $3$, denote the students and let the shares of time and cupboard be the first issue and the second issue respectively. We let $x_{i,k}$ denote the $i^{th}$ player's proposal for the $k^{th}$ issue and let $x_{i,k} \in \{0.1, \ldots, 0.9\}$ for simplicity reasons where $i = 1, 2, 3$ and $k = 1, 2$. We assume that there are at most three negotiation rounds and the bidding orders are $\langle (1, 2, 3), (2, 3, 1), (3, 1, 2), \ldots \rangle$. We define the following valuation functions for an agreement $\mathbf{a} = (\mathbf{x}_1, \mathbf{x}_2, \mathbf{x}_3)$ for each of the players:

$$v_1(\mathbf{x}_1) = 8 \times x_{1,1} + 2 \times x_{1,2} - 1.5$$

$$v_2(\mathbf{x}_2) = 5 \times x_{2,1} + 5 \times x_{2,2} - 1.3$$

$$v_3(\mathbf{x}_3) = 3 \times x_{3,1} + 7 \times x_{3,2} - 1.4$$

These valuation functions are part of the utility function of each player $i$, as defined in Equation (1). We set the discount factor to $\delta = 0.8$. The optimal bid of every player in every round according to the equilibrium strategy is given in Table 1 below. This table shows that in each round, the three bids form an agreement, and that every player's utility is at least as high as his utility in the next round. Unless one or more players submit bids other than their equilibrium strategies, the negotiation will stop at the end of the first round.

As a final example, suppose that player 3 bids $(0.8, 0.8)$ in the last round, deviating from its equilibrium strategy. In that case, either player 1 or player 2 will not receive enough of the issues to obtain a positive utility. Therefore, no agreement will be reached and every player will get zero utility.

**Table 1.** Example of 3-player 2-issue Negotiation

| round 1 | $\mathbf{x}_i^1$ | $u_i^1$ |
|---|---|---|
| $i = 1$ | $(0.2,\ 0.1)$ | $0.3 > 0.24$ |
| $i = 2$ | $(0.7,\ 0.2)$ | $3.2 > 2.16$ |
| $i = 3$ | $(0.1,\ 0.7)$ | $3.8 > 3.6$ |
| round 2 | $\mathbf{x}_i^2$ | $u_i^2$ |
| $i = 2$ | $(0.7,\ 0.1)$ | $2.16 > 0.128$ |
| $i = 3$ | $(0.1,\ 0.8)$ | $3.6 > 3.584$ |
| $i = 1$ | $(0.2,\ 0.1)$ | $0.24 > 0.192$ |
| round $\gamma$ | $\mathbf{x}_i^\gamma$ | $u_i^\gamma$ |
| $i = 3$ | $(0.7,\ 0.7)$ | $3.584 > 0$ |
| $i = 1$ | $(0.2,\ 0.1)$ | $0.192 > 0$ |
| $i = 2$ | $(0.1,\ 0.2)$ | $0.128 > 0$ |

## 4 Related Work

In this section, we discuss some work related to multi-player and/or multi-issue negotiation with complete information. The best known negotiation model is the alternating-offer bargaining game [14]. Basically, in a two-player game, one player proposes a partition of a single issue to the other player. The opponent can accept the proposal, make a counter-proposal, or quit the negotiation. The negotiation continues until reaching an agreement or a finite deadline. Ståhl identifies the optimal strategies for rational players with perfect information in such a game with a finite time horizon [17]. Rubinstein identifies a unique SPE, which is reached immediately, in a perfect information setting with an infinite time horizon [14]. The Ståhl-Rubinstein model [14], two-player single-issue bargaining, has been extended into two directions, either the negotiation between more than two players or the multi-issue negotiation. The model of $n$-player single-issue negotiation has been investigated in [2, 9, 16]. One proposer is chosen by a pre-specified order in each stage of a multi-stage game; he proposes a partition of one issue for all players and other players then respond sequentially. If players have time preferences with a common constant discount factor, there is a unique allocation of a pie amongst three players, which tends to an equal partition as players become more patient [9]. It is possible to obtain an equilibrium similar to the unique SPE of the two-player game by limiting the strategies available to players to history independent strategies [16]. Some other work addresses the multiple players game by modifying the structure of the game. For instance, players are engaged in a series of bilateral negotiation [8] and any player that reaches a satisfactory agreement may "exit" the game [4, 11]. Two-player multi-issue negotiation has been studied in two ways. Multiple issues are negotiated one by one, so the role of a negotiation agenda has been studied by various work [1, 3, 5, 7, 10]. Alternatively, multiple issues can be treated as one package. A comparison between the package approach and the sequential approach is made in [6] and the former shows a better outcome as it introduces the opportunity of trade-offs.

The model built in this work includes both many players and many issues. We let each player bid only for himself sequentially; every bid is searched in an inherently infinite set of bids. The game is multilateral and all issues are negotiated as one package. Both a common deadline and a common constant discount factor are set; players are not permitted to exit. A model of many player and multidimensional issue spaces has been studied in [13]. In that work, according to a pre-specified vector of "access probabilities", one proposer is selected in each negotiation round. The solution is a limit of equilibrium outcomes, as the number of negotiation rounds increases without bound. Their model let $n$ players form multiple admissible coalitions. If an admissible coalition has the proposer and his proposal is accepted by all members in that coalition, the proposal will be imposed to all $n$ players. The model is more practical, especially in political field. Compared to it, our model is more general and can be directly used on $n$ equal players.

## 5 Conclusions and Future Work

In this paper, we have proposed a general multi-player multi-issue multilateral negotiation protocol. Given two time constraints, the deadline and the discount factor, we proposed equilibrium strategies under a complete information setting. Given these strategies, an agreement can be reached immediately at the end of the first round, if it exists, and the solution is a subgame perfect equilibrium. By making trade-offs between issues, every player's utility in the equilibrium outcome is maximized and the solution is Pareto optimal. To our knowledge, this is one of the first papers to study, from a game theoretic perspective, the case of multi-issue negotiation with multiple players. This case introduces a new level of complexity to deriving subgame perfect equilibrium strategies, in comparison to bilateral bargaining. The result of this work can be widely and directly used to solve allocation problems of resources, tasks, etc.

With the technique developed in this paper, we are currently developing a solution for the incomplete information cases, in which the optimal actions of players are concerned with their beliefs about types of each other. This is a complex problem as those beliefs will change due to ongoing new bidding information.

Besides the incomplete information case, there are several other interesting directions for extending this work. It will be interesting to study a model where different players have different deadlines and discount factors also. If the bidding order of each round cannot be determined before the negotiation, the equilibrium strategies will be quite different. Finally, we can try relaxing the constraint of monotonicity and study the model with more general utility functions.

## Acknowledgements

This research has been partially funded by the Netherlands Organization for Scientific Research (NWO) under the project of Task Coordination for Non-cooperative Agents (TaCo).

## References

1. Bac, M., Raff, H.: Issue-by-Issue Negotiations: The Role of Information and Time Preference. Games and Economic Behavior 13(1), 125–134 (1996)
2. Binmore, K.: Bargaining and Coalitions. Game-Theoretic Models of Bargaining, 269–304 (1985)
3. Busch, L.A., Horstmann, I.J.: Bargaining Frictions, Bargaining Procedures and Implied Costs in Multiple-Issue Bargaining. Economica 65(256), 669–680 (1990)
4. Chae, S., Yang, J.A.: An N-person pure bargaining game. Journal of Economic Theory 62(1), 86–102 (1994)
5. Fatima, S.S., Wooldridge, M., Jennings, N.R.: An agenda-based framework for multi-issue negotiation. Artificial Intelligence 152(1), 1–45 (2004)
6. Fatima, S.S., Wooldridge, M., Jennings, N.R.: Multi-Issue Negotiation with Deadlines. Journal of Artificial Intelligence Research 27, 381–417 (2006)
7. Fershtman, C.: The importance of the agenda in bargaining. Games and Economic Behavior 2(3), 224–238 (1990)
8. Gerding, E.H., Somefun, D.J.A., La Poutre, J.A.H.: Multi-attribute Bilateral Bargaining in a One-to-Many Setting. In: Faratin, P., Rodríguez-Aguilar, J.-A. (eds.) AMEC 2004. LNCS (LNAI), vol. 3435, p. 129. Springer, Heidelberg (2006)
9. Herrero, M.: A Strategic Bargaining Approach to Market Institutions. Unpublished Ph. D. Thesis. University of London (1985)
10. Inderst, R.: Multi-issue Bargaining with Endogenous Agenda. Games and Economic Behavior 30(1), 64–82 (2000)
11. Jun, B.H.: A Structural Consideration on 3-Person Bargaining. Chapter III in Essays on Topics in Economic Theory, Ph.D. Thesis, Department of Economics, University of Pennsylvania (1987)
12. Osborne, M.J.: An introduction to game theory. Oxford University Press, New York (2004)
13. Rausser, G., Simon, L.: A non-cooperative model of collective decision making: A multilateral bargaining approach. Department of Agricultural and Resource Economics, University of California, Berkeley (1992)
14. Rubinstein, A.: Perfect Equilibrium in a Bargaining Model. Econometrica 50(1), 97–110 (1982)

15. Selten, R.: Spieltheoretische Behandlung eines Oligopolmodells mit Nachfragetragheit. Zeitschrift fur die gesamte Staatswissenschaft 12, 301–324 (1965)
16. Shaked, A.: A Three-Person Unanimity Game. Talk given at the Los Angeles national meetings of the Institute of Management Sciences and the Operations Research Society of America (1986)
17. Ståhl, I.: Bargaining theory. Ekonomiska forskningsinstitutet vid Handelshögskolan i Stockholm, EFI (1972)

# Automated Negotiation through a Cooperative-Competitive Model

Xuehong Tao[1], Zhiqi Shen[2], Chunyan Miao[3], Yin-Leng Theng[4], Yuan Miao[5], and Han Yu[6]

[1] School of Information and Communications Technology, Victoria University, Australia
Xuehong.tao@vu.edu.au
[2] School of Electrical and Electronic Engineering, Nanyang Technological University, Singapore
zqshen@ntu.edu.sg
[3] School of Computer Engineering, Nanyang Technological University, Singapore
ascymiao@ntu.edu.sg, yuhan@pmail.ntu.edu.sg
[4] School of Communication and Information, Nanyang Technological University, Singapore
tyltheng@ntu.edu.sg
[5] School of Engineering and Science, Victoria University, Australia
tyltheng@ntu.edu.sg
[6] School of Computer Engineering, Nanyang Technological University, Singapore
Yuan.miao@vu.edu.au

**Summary.** Automated negotiation has become increasingly important and pervasive since the advent of e-Business. It frees people from tedious interactions, improves the efficiency of e-business and ensures the accuracy of complex service composition. However, there are limitations of the existing negotiation models. Firstly, the majority of existing negotiation models are "price" bargain type of negotiation. It does not consider the reasons lead to the bargain position. Secondly, a few interest based negotiation models proposed in recent years are able to consider the underlying reasons of the counter party's position, therefore, have more chance to reach an agreement. However, they focus on individual's alternative solution seeking. None of these models promote the most productive human negotiation approach, especially in the global economic context, to constructively cooperate and seek for possible win-win situations. In an e-business environment, it would be more powerful if new services could be built on multiple parties' existing services to form a cooperative solution. This paper proposes a negotiation model to enable negotiation parties to exchange preferences and knowledge, develop optimal cooperative solutions for mutual benefits. It is a cooperative-competitive win-win strategy.

## 1 Introduction

E-business provides businesses with efficiency, cost saving and productivity. In the e-business environment, service consumers interact with service providers to receive services. However, in some cases, the service requested by the consumer can not be

T. Ito et al. (Eds.): Innovations in Agent-Based Complex Automated Negotiations, SCI 319, pp. 161–178.
springerlink.com 

fulfilled by the producer. The consumers and the producers need to negotiate their service requirements and offers. In addition, each parties wish to reach their best interests. Therefore, negotiation widely exists and it is a key activity in e-business.

Negotiation automation is to let software components to negotiate automatically on behalf of human being. Because of the large number of potential parties and various types of negotiation issues, automated negotiation becomes essential in the world of e-business. Meanwhile, automated negotiation has become an increasingly popular research topic.

Recently, agent technologies have been applied to automated negotiation. Negotiation automation can significantly reduce negotiation time (making large volumes of transactions possible in small amounts of time) and can also remove some of the reticence of humans to engage in negotiation (e.g., because of embarrassment or personality) [1], hence the formalization of negotiation has received a great deal of attention from the agent Community [2].

Traditionally, people use negotiation as a means of compromise in order to reach mutual agreement. In general, negotiation is defined as an interactive process which aims to achieve an agreement for business parties. Self interested agents work for their own goals and are competitive among each other by nature. In an e-business environment, it is also desirable for negotiation agents to have an incentive to cooperative in order to achieve efficient mutually beneficial win-win solutions. That is to say, cooperation is regarded as having the same level of importance as competition. Hence the new term coopetition is created to describe the cooperation-competition characteristics of business activities.

Most of the existing negotiation automations are "price" bargaining type of negotiation that focus on fixed bargaining positions, or simple interest based negotiation that focuses on seeking alternative solutions for individual agents to avoid conflicts. They are not focused on finding mutual gain solutions which will give negotiation parties an opportunity to plan on the whole (even if self interested) and make full use of all parties capabilities and maximize the overall benefit.

This paper proposes a knowledge based model for negotiation automation, and it tries to find optimal mutually beneficial solutions for the negotiation parties using shared knowledge of all parties. The rest of the paper is organized as follows: Section 2 introduces the negotiation strategies and related works. Section 3 introduces our cooperative-competitive negotiation model and section 4 proposes the computational model for negotiation agents. Section 5 provides the algorithms to automate the key negotiation processes, and Section 6 illustrates the method with a case study. Section 7 concludes the paper.

## 2 Negotiation Strategies and Related Works

**Negotiation strategies:** The traditional negotiation focuses on bargaining positions, such as price, delivery time and quantity etc. It is termed *Position Based Negotiation* (PBN). In a PBN, the involved parties are firmly committed to their bargaining positions. A position tells others what a negotiation party wants, and reflects his/her point of view on a certain issue. It does not tell others about the complex decision making process that lead to the position; nor does it provide others the opportunity to take his/her interests into account. In position based negotiation, the involved parties argue only their positions. Their underlying interests may never be explicitly mentioned. If no agreement on the positions can be reached, the negotiation fails.

An example for setting up an educational game environment is illustrated below and no agreement was made in this case

$A$ : *Could you help me develop an educational game for Primary One math class?*
$B$ : *Sorry, we don't have software development services.*
$A$ : *That is OK. Bye.*

*Interest Based Negotiation* (IBN) [3] focuses on satisfying the underlying reasons rather than to meet the stated demands. Interests of a negotiation party tell others why he/she wants something. They reflect his/her underlying concern, needs or desire behind an issue. In interest based negotiation, the interests of participants are identified and explored, which helps them to understand the others' perspectives instead of simply reacting to the positions. By discussing the reasons behind the positions and thinking of alternatives, mutually acceptable agreement is more likely to be reached.

In the above scenario, the goal of A is to set up an educational game environment, so B proposes an alternative solution: buy the game system instead of developing the game system. There will be an agreement if it is acceptable to A, see example below.

$A$ : *Could you help me develop an educational game for Primary One math class?*
$B$ : *Sorry, we don't have software development services.*
$A$ : *I want to set up an educational game environment.*
$B$ : *Do you want Math Discovery Educational Game System which is an integration of hardware, communication software and Math Discovery game software? We have it in stock.*
$A$ : *That is perfect.*

**Related work in agent community:** Intelligent agent, as a new type of autonomous components for constructing open, complex and dynamic systems, is one of the most suitable software entities to carry out negotiation automation. Agent community also takes negotiation as a core part of agent interactions. Jennings et al. [2] defined negotiation as the process by which a group of agents try to come to a mutually acceptable agreement on some matter.

The research of negotiation automation in software agent community can be categorized into three main approaches [2]: game theoretic approach [4], heuristic approach [5] and argumentation-based approach [6][7]. The game theoretic approach applies game theory techniques to find dominant strategies for each participant. The heuristic-based approach applies heuristic decision making during the course of the negotiation. Negotiators are not allowed to exchange additional information other than the proposal in both approaches. They are mainly used for position based negotiations.

The Argumentation-Based approach allows agents to exchange additional information. It enables agents to gain a wider understanding of their counterparts, thereby make it easier to resolve certain conflicts especially for conflicts due to incomplete knowledge. Argumentation based negotiation is a broad term, it refers to all the negotiations that exchange additional meta-level information (arguments) during the negotiation process [2]. This approach provides support for interest based negotiation strategy, as negotiators can exchange their pursuing interest/goals through argumentation.

There are some recent studies using argumentation based agent approach to realize interest based negotiation strategy. To list a few, Rahwan et al [8] proposed a framework for intelligent agents to conduct interest based negotiation. They studied the relationships between agent's goals and the types of arguments that may influence other agents' decisions, as well as defined a set of locutions that can be used in the negotiation procedure. Pasquier [9] gave a fully computational specification of negotiation agents using the 3APL agent language, where the agents are able to propose alternative plan(s) for the underlying goals. Tao et al designed a computational model and algorithms to fully automate the key components of interest based negotiation [10]. Based on suitable knowledge models, automated interest based negotiation is also applied in educational contexts for curriculum negotiation [11][12]. Pasquier et al [13] conducted empirical study on interest based bargaining and reframing agents, where the agents can exchange information about their underlying interests and alternatives to achieve the interests. The simulation demonstrated the advantages of interest based negotiation.

## 3 Cooperative-Competitive Negotiation Model

*Cooperative-competitive Negotiation (coopetitive negotiation)* is a new model of negotiation we propose in this paper that the negotiation parties can cooperatively use their knowledge to jointly create a solution acceptable to both parties. They can share information to have a more globalized view, they can exchange goals to pursue mutual benefits and share capabilities to develop cooperative solutions. Meanwhile, self interested agents work on their own benefits. They are competitive among each other. This model enables negotiators to find optimal solutions among competitive options. Hence, this is a new model of negotiation and it advanced interest based negotiation by introducing the cooperative-competitive characteristics. For the same educational game set up scenario, better solutions could be developed if it is based on multiple parties' knowledge and capabilities. As illustrated in the following example, a solution could be using the existing hardware and buying software from B with a total cost $5000, or a more cost effective solution to buy software from C and B respectively with a total cost $4500.

$A$ : *Could you help me develop an educational game for Primary One math class?*
$B$ : *Sorry, we don't have software development services.*
$A$ : *I want to set up an educational game environment.*

$B$ : *Do you want Math Discovery Educational Game System for \$8000 which is an integration of hardware, communication software and Math Discovery game software? We have it in stock.*
$A$ : *I already have our hardware system.*
$B$ : *You can use your hardware system and buy communication software (\$2000) and Math Discovery game software (\$3000) from here.*
$A$ : *Ok, the total cost is \$5000 and that is good.*

Or another party involve in the negotiation:

$A$ : *I already have our hardware system.*
$B$ : *You can use your hardware system and buy communication software (\$2000) and Math Discovery game software (\$3000) from here.*
$C$ : *I sell communication software for \$1500. You can use your hardware, buy communication software from me, and buy Math Discovery game software from B.*
$A$ : *Ok, the total cost is \$4500 and that is excellent.*

This example demonstrated that a good negotiation strategy should exhibit the following capabilities:

- Finding alternative solutions when no agreement on stated positions.
- Exchanging information to form a globalized view.
- Choosing the optimal among competitive solutions.
- Seeking cooperative solutions that aggregate individual's capabilities.
- Pursuing mutual benefits which form the foundation of long term cooperation.

Unlike the existing interest based negotiation models, the proposed model not only uses the argumentation based approach to exchange goals or preferences, provides alternative solutions to avoid conflicts but also introduces cooperative and competitive characteristics.

More specifically, our model distinguishes itself from the existing interest based negotiation in the following aspects. Firstly, most of the existing interest based negotiation models focus on individual alternative solution seeking so as to avoid conflicts. Our model focuses on multiple party joint solution construction to resolve the conflicts. It is a cooperative solution. Secondly, the methods in existing interest based negotiations are to find a solution without conflict. Our model is able to find the optimal solution during the process of searching for nonconflicting solutions. It is a competitive solution. Thirdly, some existing methods have restrictions that higher level goals (from the same agent or different agents) cannot share sub goals or resources, so as to remove the potential conflict. They are more suitable to model agents that work separately and in separate domains. Our model allows agents to share sub-goals and resources, and enables agents (even if self interested) to build solutions that satisfy the combined goals from multiple parties. Overall, our model advances the existing interest based negotiation methods by introducing the cooperative competitive characteristics.

# 4 Cooperative-Competitive Negotiation Agent

## 4.1 Overview

Agents are autonomous entities that make decision independently and work towards their goals. Complex goals can be considered as a composition of sub goals. Sub goals may be further decomposed to next level sub goals. The goals and the sub goals form a hierarchical structure. The goals and their relationships are the knowledge of agents to interact with the environment and evolve. The knowledge is maintained in the knowledge base of agent.

The Coopetitive Negotiation Agent proposed in this paper is a generic model representing the core parts of cooperative-competitive negotiation. The main components are a knowledge base and a negotiation engine.

The knowledge base stores the knowledge about goals. The negotiation engine manages the negotiation process and generates negotiation solutions automatically. It has the following main functionalities:

- Generate a Proposal: In the context of e-Business context, for service provider, a proposal is an offer to consumers for certain services. For service consumer, a proposal is a request for certain services.
- Accept/Reject a Proposal: Whether to accept or to reject a proposal depends on many factors, including whether a consumer needs the offer, whether the provider is able to offer the service and whether the price, time, quality or other criteria are satisfied.
- Exchange Information: An agent normally has incomplete knowledge. So the decision is made based on limited local information. If agents exchange information during the negotiation, it is possible to find more options for solving a problem. Hence there are more chances to achieve an agreement.
- Develop a Mutual Beneficial Solution: Agents have the ability to make use of information shared from other agents, find a solution to meet goals of all agents.
- Alter Negotiation Positions: If no agreed deal is reached, an agent may consider changing to other sub goal(s) while still supporting the same super goal.

## 4.2 Knowledge Model for Cooperative-Competitive Negotiation Agent

In e-business environment, a negotiation agent should have knowledge about its goals and how complex goals can be composed from elementary goals where the elementary goals can be achieved by primary services. The knowledge base of a negotiation agent is a collection of goals and relationships among goals. It is defined as a 3-tuple $KB$ = <$G$, $R$, $C$>, where

$$G = \{g_i = 1, 2, \cdots n.\}$$
$$R = \{r_i : g_{i0} \rightarrow g_{i1}, g_{i2}, \cdots g_{ik} | g_{i0}, g_{i1}, \cdots g_{ik} \in G, i = 1, 2, \cdots m\}$$
$$C = \{c(g) | g \in G\}$$

$G$ is a goal set, $R$ is a relationship set where each relationship $r_i$ describes how a super goal is decomposed to sub goals. $g_{i0}$ is termed as the head of a relationship, $g_{i1}, g_{i2}, \cdots g_{ik}$ are termed as the tail of a relationship.

$C$ is a criteria set which will be discussed later. $c(g)$ is the criteria values relevant to $g$, such as price, delivery time, quality of service, payment methods and etc.

According to the super-sub goal relationship, goals of an agent form a goal hierarchy, which is a network and it is not necessary a tree.

- **Atom Goal**

A goal $g$ is called an atom goal if there is no decomposition relationship such that it has $g$ as the head and other goals as the tail. Atom goals are goals that can not be decomposed to other sub goals. They are corresponding to the primary services in an agent's belief.

An atom goal of one agent maybe a composite goal of another agent, because agents have different belief about the basic services they can operate. For example, for a real estate agent, obtaining a house is an atom goal. However it is a composite goal for a builder agent which may contains a sub goal of buying a block of land and a sub goal of building a house.

- **Decomposition**

Following some relationship in $R$, a goal $g$ can be decomposed into sub goals (not necessarily atom goals). The set of the sub goals are called a decomposition of $g$. A goal may have different decompositions.

A goal is achievable if it can be decomposed to a set of atom goals, and the services corresponding to the atom goals are all available.

For example, in a holiday booking scenario,

$$
\begin{aligned}
G = \{ & g_1 = \text{"have holiday booking"}, \\
& g_2 = \text{"have transport booking"}, \\
& g_3 = \text{"have accommodation booking"}, \\
& g_4 = \text{"obtain air ticket from X Airline"}, \\
& g_5 = \text{"obtain booking of A Hotel"}, \\
& g_6 = \text{"obtain train ticket from Y railway services"} \} \\
R = \{ & r_1 : g_1 \rightarrow g_2, g_3, r_2 : g_2 \rightarrow g_4, r_3 : g_3 \rightarrow g_5, r_4 : g_2 \rightarrow g_6 \}
\end{aligned}
$$

Here, $\{g_2, g_3\}$, $\{g_4, g_3\}$, $\{g_4, g_5\}$ and $\{g_6, g_5\}$ are all decompositions of $g_1$. Goal $g_1$ can be achieved by $\{g_4, g_5\}$ or $\{g_6, g_5\}$ , i.e. for a holiday booking, one solution is to take flight of airline X and live in Hotel A. Another solution is to go by train from Y Railway services and live in Hotel A.

- **Criteria of Goals**

There are some criteria to describe a goal (service), such as price, delivery time, quality of service, payment methods and etc. We define the criteria of a goal $g$ as a vector $(v_1, v_2, \cdots, v_n)$ from a domain vector $(D_1, D_2, \cdots, D_n)$.

$c(g) = (v_1, v_2, \cdots, v_n) \in (D_1, D_2, \cdots, D_n)$, $D_i$ is the domain of $v_i$.

For example, if a goal $g$ is "Buying a Lenovo Notebook model S10". $c(g)$=($900, 2, {cash, credit card}) from domain ($R^+$, $I^+$, {cash, credit card, bank transfer}). This may mean, the price is $900 from a positive real number domain, the delivery time is two days from a positive integer domain, and the payment method is either by cash or by credit card from a set domain contains all possible payment methods.

For a certain service, the values in the criteria allow negotiators to make comparison between competitive solutions and to request an optimal one. Suppose agents are able to compare the preference among multi-criteria[14]. For example a simple way could be by using weight to combine all dimensions in the criteria to a single value then compare this single value.

In the rest of this paper, we consider criteria as a single value and suppose the smaller value is the better without loss of generality. For composite goals, they have different decompositions each having different criteria values. $c(g)$ is the smallest among them or a lower bound of them. The estimated criteria of composite goals can be used as a heuristic in search algorithms. Choosing a small estimated value can make sure the goal has more opportunity to be considered. For atom goals, if it corresponds to an available service, $c(g)$ is the actual service criteria value. If it is corresponding to an unavailable service according to the agent's knowledge, $c(g) = +\infty$.

- **AND/OR Graph Representation of Knowledge Base**

For easy presentation of our algorithms, we also define the graph representation of a knowledge base. An AND/OR Graph[15] is a hyper graph. Instead of arcs connecting pairs of nodes in the graph, there are hyper arcs connecting a parent node with a set of successor nodes. These hyper arcs are called connectors. Suppose $KB$=$<G_{KB}, R, C>$ and its AND/OR Graph representation is $Q = (G_Q, E, C)$, where

$G_Q = G_{KB}$, i.e. nodes in $Q$ are the goals in $KB$,
$E = \{(g_{i0}, \{g_{i1}, g_{i2}, \cdots g_{ik}\}) | g_{i0} \rightarrow g_{i1}, g_{i2}, \cdots g_{ik} \in R\}$, i.e. connectors in $Q$ are decomposition rules in $KB$.

Leaf nodes in $Q$ are atom goals in $KB$.

- **Solution Graph and Partial Solution Graph**

In an AND/OR graph $Q$, a node $g$ can be expanded to its successors by following exactly one connector. Each successor node can be expanded further in the same way and a graph rooted on $g$ will be generated. The graph is called a *Partial Solution Graph* of $g$. If all the leaves of the partial solution graph are the leaves of $Q$, the partial solution graph is a solution graph. Partial solution graph and solution graph are graph representations of goal decompositions.

In the above holiday booking example, the AND/OR Graph representation of the knowledge base is shown in Figure 1(a). Two possible solution graphs are shown in Figure 1(b) and two partial solution graphs are shown in Figure 1(c).

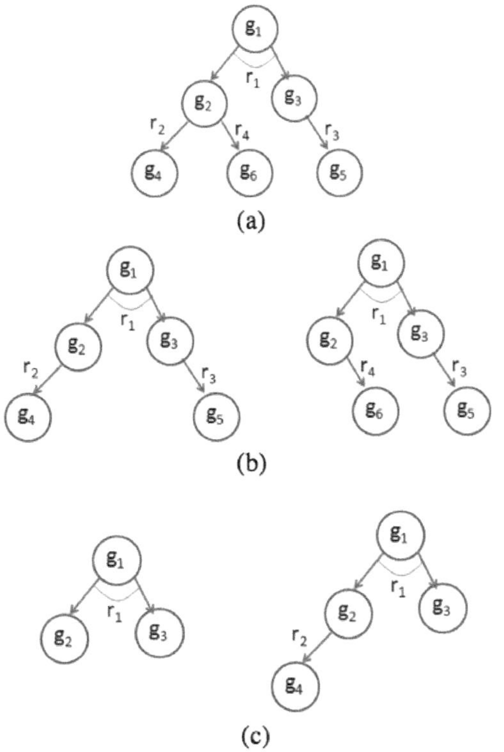


**Fig. 1.** Graph representation of the holiday booking $KB$. (a) Graph representation of the knowledge base (b) Possible solution graphs (c) Partial solution graphs

Suppose the knowledge base of an agent is maintained periodically so that it has no loop decomposition and the decompositions are all minimal. The requirement of non loop decomposition means a goal's decomposition can not include the goal itself. Formally, there is no decomposition $Z$ of a goal $g$ such that $g \in Z$. Minimal decomposition means there is no decompositions $Z_1$ and $Z_2$ of a goal $g$ such that $Z_1 \subset Z_2$. i.e. the rules will not produce unnecessary sub goals. For example, if $\{g_1, g_2\}$ and $\{g_1, g_2, g_3\}$ are two of the decompositions of a goal, then it does not meet the minimal decomposition requirement because $g_3$ is unnecessary.

Knowledge Base Revision [16] can provide the system with learning capabilities by adding in new knowledge and removing/revising existing knowledge during the negotiation process. The details of knowledge base revision will be omitted here.

## 5 Negotiation Automation

- **Goal Decomposition Algorithm**

Firstly, we will provide a method to decompose a goal, named $g$, to atom goals (which correspond to primary services) using a heuristic search strategy.

Suppose we have a knowledge base $KB$ which contains relationships about goal decompositions. For an atom goal, if it corresponds to an available service, $c(g)$ is the actual service criteria value. If it is not available, $c(g) = +\infty$. Suppose the agent is able to perform multiple criteria preference analysis [14] and find the solution with the optimal criteria. For simplicity, we consider the smaller criteria solution as the better one.

Algorithm Decompose listed below will decompose $g$ to atom goals based on Nilsson's AO* algorithm [15]. During the process of creating a search graph and marking a partial solution graph, the algorithm is gradually approaching to the optimal solution by using the criteria of each goal as heuristics. The algorithm starts from $g$, selects and marks the connector with the smallest criteria as the temporary best solution for $g$. Then continues to decompose the sub-goals of $g$. Whenever new information that makes changes to the criteria of a goal is encountered, the algorithm will propagate the newly discovered information up the goal hierarchy, re-calculate the criteria and make a new selection among connectors.

**Algorithm. Decompose** $(g)$

1. Create a search graph $Q, Q = \{g\}$
   If $g$ is an atom goal, label $g$ as *Solved*. cost $(g)$ = c $(g)$
2. Until $g$ is labeled *Solved* , or cost $(g) = +\infty$ do
   a. // Select node to expand Compute a partial solution graph H in $Q$ by tracing down marked connectors in $Q$ from $g$ (marks will be discussed later in this algorithm)
      Select any non terminal leaf node $n$ of H
   b. // Expand node $n$ by generating its successors
      - If $n \rightarrow n_1, n_2 \cdots n_k \in R$, Add all sub goals of $n$ to $Q$
      - For successors $n_j$ not occurring in $Q$, cost($n_j$)= c($n_j$)
      - If $n_j$ is leaf, label *Solved*.
   c. // Propagate the newly discovered information up the graph
      $S = \{n\}$ // $S$ is a set of nodes that have been labeled
      // solved or whose cost have been changed

      Until $S$ is empty do
      - Remove a node $m$ ( $m$ has no descendants in $S$) from $S$
      - //Computer the cost of each $m$'s decomposition
        // cost $(m)$ is the minimum cost among all connectors For each connector $m \rightarrow m_{i1}, m_{i2}, \cdots, m_{ik}$
        $Cost_i(m) = cost(m_{i1}) + cost(m_{i2}) + \cdots + cost(m_{ik})$
        $Cost(m) = min_i(cost_i(m))$
        Mark the best path out of $m$ by marking the connector with minimum cost
      - If all nodes connected to $m$ through this new marked connector has been labeled *solved*, label $m$ *solved*
        If $m$ solved or cost of $m$ just changed, add all of the ancestors of $m$ to $S$
3. If $g$ is labeled *Solved*, return *True*, else return *False*

**End of Decompose.**

- **Proposal Generation**

An agent selects its high level goal, named $g$ based on certain reasoning mechanism. If algorithm Decompose ($g$) returns *True*, the partial solution graph H is the current pursuing solution graph for goal $g$. Based on H, if a goal can not be realized by the agent itself, it will be proposed to other agents. A proposal could be an offer proposal from the provider agent to advise its services, or a request proposal from the consumer agent to ask for services.

Hence a proposal is a goal $g \in$ H. It can be an atom goal for a single service, or a composite goal for a complex service.

- **Cooperative-Competitive Solution Construction**

When an agent receives a proposal $g$, it will evaluate it and then decide whether to accept or deny it. If no agreement can be reached, the participating agents may consider exchanging negotiation related information, including information from $KB$ and pursuing goals.

Upon receiving new knowledge from other agent(s), the agent will carry out a temporary knowledge base revision by adding the new knowledge to its existing knowledge base. Whether to incorporate the new knowledge permanently in the knowledge base will be decided by the agent through other mechanism. The temporary knowledge base revision can be implemented by algorithm KBRevision listed below.

Suppose the knowledge base of the agent is $KB$=<$G$, $R$, $C$>, and the agent will revise the $KB$ to incorporate new knowledge noted as $KB$'=< $G$',$R$',$C$'>.

**Algorithm. KBRevision( )**
For each new goal in G', add into G
For each new relationship in R', add into R if it doesn't cause loop decomposition
For each new criteria $c_{new}(n)$
If there is no criteria of $n$ exists in $KB$, add $c_{new}(n)$ into C
If there is criteria $c_{old}(n)$ exists and $c_{old}(n) \neq c_{new}(n)$,

a. $c(n) = min(c_{old}(n), c_{new}(n))$, which makes sure the low criteria solution has the opportunity to be selected
b. propagate the new criteria to upper lever goals (details will be omitted here as it is similar as what have been done in algorithm Decompose, step 2.c.)
c. If $n$ is an atom in $KB$'
Add $n \rightarrow n$' in $KB$, c($n$')=$c_{new}(n)$
If $n$ is an atom in $KB$
Add $n \rightarrow n$" in $KB$, c($n$")=$c_{old}(n)$

**End of KBRevision.**

Based on the newly build temporary knowledge base,
If Decompose ($g$) =*True*
Partial solution graph H is the solution to $g$

This solution is a cooperative solution because it is constructed on both parties' available options. It is also a competitive solution because it selected the best cost solution.

- **Mutual Beneficial Solution Construction**

If party A has goals $g_A^1, g_A^2, \cdots g_A^s$ and party B has goals $g_B^1, g_B^2, \cdots g_B^t$, they want to seek opportunity to achieve their mutual goals. We can add decomposition knowledge $g_{\text{Mutual}} \rightarrow g_A^1, g_A^2, \cdots g_A^s, g_B^1, g_B^2, \cdots g_B^t$, into the knowledge base. If Decompose ($g_{\text{Mutual}}$) is $True$, the partial solution graph $H$ is the solution to $g_{\text{Mutual}}$.

- **Negotiation Position Alternation**

If there is no solution for the current proposal $g$, the participating agents may also consider other alternative goals that support the same super goal as that $g$ does. This can be achieved by

$f$ = father of $g$ in the current pursuing solution graph G
make $f$ the new proposal

By doing so, the agent changes the negotiation position from $g$ to $f$, and work on other possibilities to achieve $f$.

- **Correctness and Advantages of the Method**

If no solution for $g$, i.e. all decompositions of $g$ contain unavailable services, according to the algorithm cost($g$) will reach $+\infty$ , so the algorithm returns false.

If there is a solution from $g$ to a set of atom goals, and if for all goal decomposition relationship $n \rightarrow n_1, n_2 \cdots n_k, c(n) \leq c(n_1)+c(n_2) \cdots + c(n_k)$, the algorithm will terminate and return $True$. By tracing the marks, graph H is the optimal solution. cost($g$) is the cost of the solution.

Hence, with the restriction that for all composite goal $g$, the estimated criteria c($g$) is always smaller than the sum of its sub goals, i.e. the estimated criteria is always smaller than the real criteria, the algorithm can find the optimal solution.

By limiting the estimated criteria of a goal $g$ to be not bigger than the actual criteria, the actual low criteria solution of $g$ will have the opportunity to be explored. However, if the estimated criteria are much lower than the actual criteria, this will direct the algorithm to spend time to explore this seemingly optimal but actually not optimal branch. Hence a good estimation will reduce the unnecessary search and find the optimal solution.

The proposed method is flexible in handling negotiation conflicts and has the following advantages:

- Find alternative solutions or alter pursuing goals when there is no agreement on initial negotiation positions.
- Find cooperative solutions based on the knowledge of multiple parties.
- Find optimal solutions among competitive options.
- Find mutual beneficial solutions by using a joint goal.

# 6 Case Study

## 6.1 The Conference Management Problem

AB University (ABU) wants to organize a conference. Some conference matters planned include arranging meeting rooms, printing meeting materials and organizing a museum visit activity. ABU hopes to reduce the cost so it is going to use its own meeting room (cost \$0), print meeting materials by itself (cost \$3000) and ask a third party to arrange the museum visit (with estimated cost \$5000). So the planned budget is \$8000.

XY Event Management Company (XYE) can provide conference management services for others include rent rooms, print meeting materials and organize a city tour. It also organizes celebration activities such as birthday parties, weddings.

The negotiation starts when AB University requests museum visit event management from XY Event Management.

**Position Based Negotiation:** In position based negotiation, the negotiation will stop when XYE replies that it doesn't provide museum visit event management.

**Interest Based Negotiation:** Considering interest based negotiation, ABU and XYE may further discuss why museum visit is needed. After knowing museum visit is part of the entertainment activity of a conference, XYE is likely to suggest replacing the museum visit by a city tour and propose to provide the whole conference management service to ABU, with total cost \$10000. The cost includes room rental \$3000 from CD Hotel, \$4000 for meeting material printing and \$3000 for the city tour. If ABU accepts the proposal, the negotiation may stop here.

## 6.2 Cooperative-Competitive Negotiation

In real world, ABU and XYE may further seek cooperation to optimize the solution. For example, with city tour to be arranged by XYE, and conference to be hosted in ABU's conference rooms instead of spending the cost to rent rooms from CD Hotel. Now let's use our proposed Cooperative-Competitive model to illustrate the automated negotiation processes.

Suppose the agent $A_1$ of ABU negotiates with the agent $A_2$ of XYE for relevant services. For simplicity of presentation, we define some symbols to represent the goals. Suppose

$g_1$ : Organize conference
$g_2$ : Self-organize the conference
$g_3$ : Arrange meeting room
$g_4$ : Print meeting materials
$g_5$ : Arrange museum visit
$g_6$ : Rent room from CD Hotel
$g_7$ : Use AB University meeting room
$g_8$ : Operate business

$g_9$ : Out source conference management (ABU) / Provide conference management for others (XYE)
$g_{10}$ : Manage celebration activity
$g_{11}$ : Arrange city tour
$g_{12}$ : Rent meeting room from EF Centre

Suppose the knowledge base of $A_1$ is $KB_1$ and the knowledge base of $A_2$ is $KB_2$. For simplicity, we put the (estimated) prices with the goals together.

$KB_1 = (G_1, R_1, C_1)$ where
$G_1 = \{g_1(\$8000), g_2(\$8000), g_3(\$0), g_4(\$3000), g_5(\$5000), g_6(\$3000), g_7(\$0)\}$
$R_1 = \{g_1 \rightarrow g_2; g_2 \rightarrow g_3, g_4, g_5; g_3 \rightarrow g_6; g_3 \rightarrow g_7; \}$

$KB_2 = (G_2, R_2, C_2)$ where
$G_2$=\{$g_1$(\$9000), $g_3$(\$3000), $g_4$(\$4000), $g_6$(\$3000), $g_8$(\$9000), $g_9$(\$10000), $g_{10}$(\$8000), $g_{11}$(\$3000), $g_{12}$(\$4000) \}
$R_2 = \{g_1 \rightarrow g_9; g_8 \rightarrow g_9; g_8 \rightarrow g_{10}; g_9 \rightarrow g_3, g_4, g_{11}; g_3 \rightarrow g_6; g_3 \rightarrow g_{12}; \}$

The current goal of $A_1$ is to "organize conference". After calling Decompose ($g_1$), the solution graph is listed in Figure 2 (the price is listed beside each goal node). $A_1$ proposes to use its meeting room with no cost ($g_7$), print meeting materials by itself ($g_4$) and request others to arrange the museum visit ($g_5$). The total cost is about \$8000.

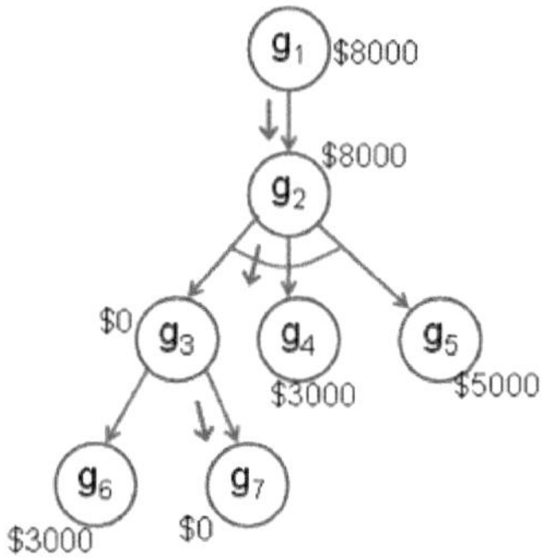


**Fig. 2.** Solution Graph of $g_1$ in $A_1$

There is no service relevant to "arrange museum visit ($g_5$)", $A_2$ rejected the proposal. No agreement on the initial proposal, $A_1$ will consider altering the initial negotiation position. $A_1$ will share its goal "self organize conference ($g_2$)". $A_2$ still has no relevant services. $A_1$ will continue to share its higher level goal "organize conference ($g_1$)".

With the knowledge that $A_1$ is aiming to organize the conference, $A_2$ knows that "organize conference" can be done by not only "self organize the conference" ($g_1 \rightarrow g_2$) but also "out source conference management" ($g_1 \rightarrow g_9$). $A_2$ is able to "provide conference management for others" ($g_9$), so it provides an alternative solution to $A_1$ that $A_2$ will help $A_1$ to organize the conference and replace the "arrange museum visit ($g_5$)" with "arrange city tour ($g_{11}$)". The solution graph (by tracing down the marks from $g_1$) is listed in Figure 3. The total cost is \$10000. Because the algorithm only expands the relevant nodes, goals such as $g_8$ and $g_{10}$ are not considered here.

With the relevant knowledge shared by $A_2$, $A_1$ could revise its knowledge base to incorporate the new knowledge $KB$'=<$G$', $R$', $C$'> contained in $A_2$'s proposal, where $G$'=$\{g_1, g_9, g_3, g_4, g_{11}\}$, $R$'=$\{g_1 \rightarrow g_9; g_9 \rightarrow g_3, g_4, g_{11};\}$, $C$'=$\{c(g_4)$=\$4000, $c(g_{11})$=\$3000, $c(g_9)$=\$10000$\}$. After using algorithm

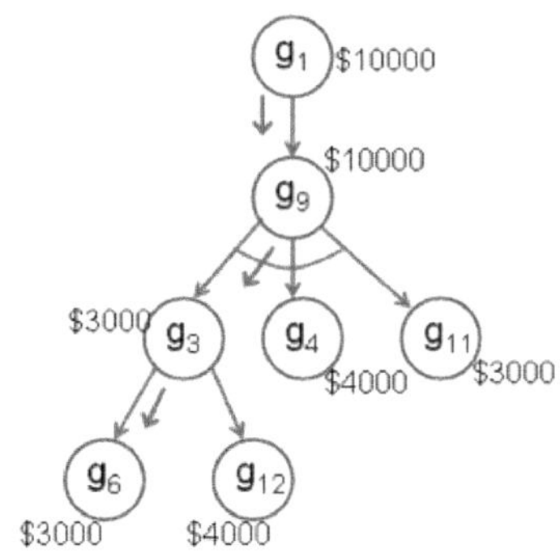


**Fig. 3.** Solution Graph of $g_1$ proposed by $A_2$

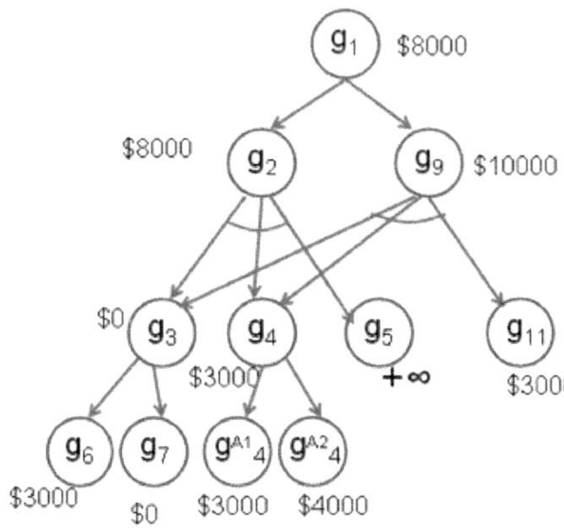


**Fig. 4.** The temporary knowledge base of $A_1$

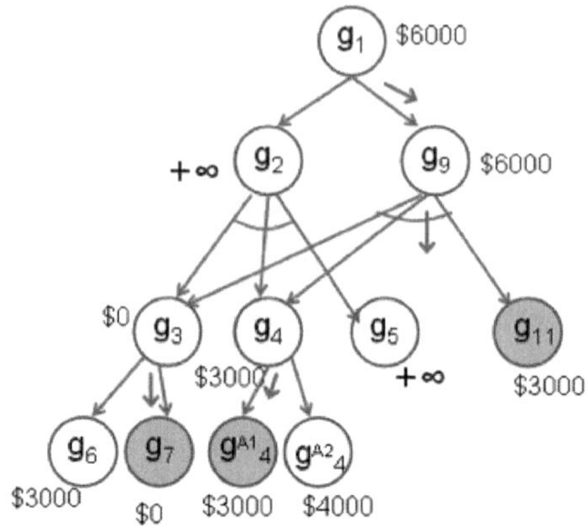


**Fig. 5.** Solution Graph of $g_1$ based on shared knowledge

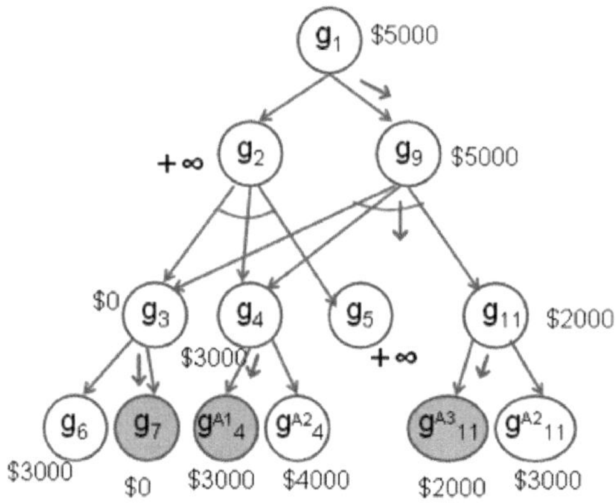


**Fig. 6.** Three Parties Cooperative Solution Graph of $g_1$

*KBRevision* to incorporate the new information, the temporary knowledge base of $A_1$ is as shown in Figure 4.

From the temporary knowledge base, $A_1$ could use Decompose ($g_1$) to build a solution graph as shown in Figure 5. The total cost is $6000.

If there is a Tourism Company (TC), whose agent $A_3$ shares knowledge about its service "arrange city tour ($g_{11}$)" with the cost of $2000, a more cost effective solution could be built as shown in Figure 6. The total cost is $5000. The final solution constructed is a cooperative solution from three parties and with the best cost among the competitive options.

As it shows, most of the current interest based negotiations focus on individual alternative solution seeking, whereas our model is able to build alternative multiparty joint solutions and choose the most effective one.

## 7 Conclusions and Future Works

This paper proposed a new computational model for negotiation automation: coopera-tive-competitive negotiation. Cooperative-competitive negotiation allows involved parties to dig into the higher level goals behind their positions and use mutual knowledge to construct new solutions, so the solutions are planned based on knowledge and preference from all parties and they are cooperated mutually beneficial decisions. The cooperative-competitive negotiation is more powerful and constructive than position based negotiation or simple alternative solution seeking kinds of interest based negotiation.

In our subsequent research, we will focus on the design of knowledge models (such as using Fuzzy Cognitive Map [17], Dynamic Cognitive Networks [18]) that better represent human negotiation processes.

## References

1. Lomuscio, A.R., Wooldridge, M., Jennings, N.R.: A Classification Scheme for Negotiation in Electronic Commerce. Int. Journal of Group Decision and Negotiation 12(1), 31–56 (2003)
2. Jennings, N.R., Faratin, P., Lomuscio, A.R., Parsons, S., Sierra, C., Wooldridge, M.: Automated negotiation: prospects, methods and challenges. Int. Journal of Group Decision and Negotiation 10(2), 199–215 (2001)
3. Fisher, R., Ury, W.: Getting to Yes: Negotiating Agreement without giving in. Penguin boks, New York (1983)
4. Brams, S.J.: Negotiation Games: Applying Game Theory to Bargaining and Arbitration. Routledge, New York (2003)
5. Faratin, P., Sierra, C., Jennings, N.R.: Using Similarity Criteria to Make Issue Trade-offs in Automated Negotiations. Journal of Artificial Intelligence 142(2), 205–237 (2002)
6. Karunatillake, N.C., Jennings, N.R., Rahwan, I., Norman, T.J.: Argument-based negotiation in a social context. In: Proceedings of the fourth international joint conference on Autonomous agents and multiagent systems, pp. 1331–1332 (2005)
7. Rahwan, I., Ramchurn, S.D., Jennings, N.R., McBurney, P., Parsons, S., Sonenberg, L.: Argumentation-based negotiation. The Knowledge Engineering Review 18(4), 343–375 (2004)
8. Rahwan, I., Sonenberg, L., Dignum, F.: On Interest based Negotiation. In: Dignum, F.P.M. (ed.) ACL 2003. LNCS (LNAI), vol. 2922, pp. 383–401. Springer, Heidelberg (2004)
9. Pasquier, P., Dignum, F., Rahwan, I., Sonenberg, L.: Interest-Based Negotiation as an Extension of Monotonic Bargaining in 3APL. In: Shi, Z.-Z., Sadananda, R. (eds.) PRIMA 2006. LNCS (LNAI), vol. 4088, pp. 327–338. Springer, Heidelberg (2006)
10. Tao, X., et al.: Interest Based Negotiation Automation. In: Huang, D.-S., Li, K., Irwin, G.W. (eds.) ICIC 2006. LNCS (LNBI), vol. 4115, pp. 211–222. Springer, Heidelberg (2006)
11. Tao, X., Yelland, N., Miao, Y.: Adaptive Learning through Interest Based Negotiation. In: The 16th International Conference on Computers in Education, ICCE 2008, Taiwan (October 2008)
12. Tao, X., Miao, Y.: Interest Based Learning Activity Negotiation. In: Proceedings of Cyberworlds 2008, HangZhou, China (September 2008)

13. Pasquier, P., Hollands, R., Dignum, F., Rahwan, I., Sonenberg, L.: An empirical study of interest-based negotiation. In: Proceedings of the ninth international conference on Electronic commerce, pp. 339–348 (2007)
14. In, H.P., Olson, D.: Requirements Negotiation Using Multi-Criteria Preference Analysis. Journal of Universal Computer Science 10(4), 306–325 (2004)
15. Nilsson, Nils, J.: Principles of Artificial Intelligence. Birkhäuser (1982)
16. Eiter, T., Gottlob, G.: On the complexity of propositional knowledge base revision, updates, and counterfactuals. Artificial Intelligence 57(2-3), 227–270 (1992)
17. Miao, C.Y., Liu, Z.Q., Miao, Y., Goh, A.: DCM: Dynamical Cognitive Multi-Agent Infrastructure for Large Decision Support System. International Journal of Fuzzy Systems 5(3), 184–193 (2003)
18. Miao, Y., Liu, Z.Q., Siew, C.K., Miao, C.Y.: Dynamic Cognitive Net-an Extension of Fuzzy Cognitive Map. IEEE Transaction on Fuzzy Systems 19(5), 760–770 (2001)

# An Approximation Method for Power Indices for Voting Games

Shaheen S. Fatima[1], Michael Wooldridge[2], and Nicholas R. Jennings[3]

[1] Department of Computer Science,
Loughborough University, Loughborough, LE11 3TU, UK
s.s.fatima@lboro.ac.uk
[2] Department of Computer Science,
University of Liverpool, Liverpool, L69 3BX, UK
mjw@liv.ac.uk
[3] School of Electronics and Computer Science,
University of Southampton, Southampton SO17 1BJ, UK
nrj@ecs.soton.ac.uk

**Summary.** The Shapley value and Banzhaf index are two well known indices for measuring the power a player has in a voting game. However, the problem of computing these indices is computationally hard. To overcome this problem, we analyze approximation methods for computing these indices. Although these methods have polynomial time complexity, finding an approximate Shapley value using them is easier than finding an approximate Banzhaf index. We also find the absolute error for the methods and show that this error for the Shapley value is lower than that for the Banzhaf index.

## 1 Introduction

Coalition formation is a key form of interaction in multi-agent systems. It is the process of joining together two or more agents so as to achieve goals that individuals on their own cannot, or to achieve them more efficiently [10]. Often, in such situations, there is more than one possible coalition, and the agents/ players must decide how to form a coalition and how to split the gains of cooperation between the members of a coalition. In this context, cooperative game theory offers a number of solution concepts such as *core*, *kernel*, and *stable solution* [10]. A number of multiagent systems researchers have used and extended these solutions to facilitate automated coalition formation [16, 17, 14, 13]. A key problem, in the context of multi-agent systems, is to study the computational aspects of the solutions that game theory provides. For example, [4] shows that the problem of finding the core is NP-complete. Another problem with these solutions is that, often there is more than one possible solution.

In order to overcome the problem of multiple solutions, Shapley proposed a solution called the Shapley value [15]. The Shapley value not only provides a *unique* solution to coalitional games but also provides a measure of how much influence or *power* a player has in determining the outcome of a game. The higher a player's

T. Ito et al. (Eds.): Innovations in Agent-Based Complex Automated Negotiations, SCI 319, pp. 179–193.
springerlink.com 

Shapley value, the more control he has in determining the outcome of a game. Thus the Shapley value can be viewed as an index for measuring the power of players in a game. Like Shapley value, the Banzhaf index [2] is another way of measuring a player's power. However, a key drawback of both these power indices is that computing them for voting games[1] is, in general, #P-complete [5, 12]. In other words, it is practically infeasible to try to compute the exact Shapley value or Banzhaf index. Hence, in order to overcome this computational complexity we present a new randomized method for finding an *approximation* for these indices.

The time complexity of the proposed approximation methods is polynomial in the number of players. Now, the quality of an approximation is evaluated in terms of its error of approximation. To this end, we find the *absolute error* for the proposed methods and show that this error for the Shapley value is lower than that for the Banzhaf index.

Although some approximation methods for the Shapley value have been proposed in the past, to our knowledge, there has been no study of their performance in terms of the approximation errors (see Section 5 for details). This paper not only provides new approximation methods, but also analyzes them in terms of their errors.

The rest of the paper is organised as follows. Section 2 provides the background to voting games and power indices. In Section 3 we present our approximation methods. In Section 4 we analyze their absolute error. Section 5 discusses related literature. Section 6 concludes.

## 2 Background

A coalitional game $\langle N, v\rangle$, consists of:

1. a finite set, $N = \{1, 2, \ldots, n\}$ of players, and
2. a function, $v$, that associates with every non-empty subset $S$ of $N$ (i.e., a *coalition*) a real number $v(S)$ the worth of $S$ that corresponds to it.

For each coalition $S$, $v(S)$ is the total payoff that is available for division among the members of $S$.

### 2.1 Weighted Voting Game

A weighted voting game $G = \langle N, v\rangle$ is a game such that [10]:

$$v(S) = \begin{cases} 1 \text{ if } w(S) \geq q \\ 0 \text{ otherwise} \end{cases}$$

for some $q \in \mathbb{R}_+$ and $w_i \in \mathbb{R}_+$, where:

$$w(S) = \sum_{i \in S} w_i$$

[1] Voting games are an important mechanism for agents to reach consensus.

for any coalition $S$. Thus $w_i$ is the number of votes that player $i$ has and $q$ is the number of votes needed to win the game (i.e., the *quota*). This game is denoted as $\langle q; w_1, \ldots, w_n \rangle$.

### 2.2 Weighted $k$-Majority Game

For the set of $n$ players, a weighted $k$-majority game $(v_1 \wedge \ldots \wedge v_k)$ is a game where $v_t = [q^t; w_1^t, \ldots, w_n^t]$, $1 \leq t \leq k$ are weighted voting games and

$$(v_1 \wedge \ldots \wedge v_k)(S) = \begin{cases} 1 \text{ if } w^t(S) \geq q^t \text{ for } 1 \leq t \leq k \\ 0 \text{ otherwise} \end{cases}$$

where $w^t(S) = \sum_{i \in S} w_i^t$.

### 2.3 Power Indices

A *power index* for a voting game is a way of measuring a player's voting power. A player's power is his ability to turn a losing coalition into a winning one. The *Shapley value* and *Banzhaf index* are examples of power indices. The Banzhaf index, in turn, has two versions: the *absolute Banzhaf index* and the *normalised Banzhaf index*. For a voting game $G = \langle N, v \rangle$, these indices are defined as follows [15, 2].

The marginal contribution of player $i$ to coalition $S$ with $i \notin S$ is a function $\Delta_i v$ that is defined as follows:

$$\Delta_i v(S) = v(S \cup \{i\}) - v(S) \tag{1}$$

This means a player's marginal contribution to a coalition $S$ is the increase in the value of $S$ as a result of $i$ joining it. A player that makes a higher marginal contribution, on average, has a higher Shapley value. Specifically, a player's Shapley value is defined in terms of its marginal contribution as follows [15]:

**Definition 1.** *The Shapley value ($\varphi_i$) of the game $\langle N, v \rangle$ for player $i$ is the average of its marginal contribution to all possible coalitions:*

$$\phi_i = \sum_{S \subset N} \frac{|S|!(n - |S| - 1)!}{n!} \times \Delta_i v(S) \tag{2}$$

Note that for a voting game ($\langle q; w_1, \ldots, w_n \rangle$), a player's marginal contribution is either zero or one. This is because the value of any coalition is either zero or one. A coalition with value zero is called a "losing coalition" and with value one a "winning coalition". If a player's entry to a coalition changes it from losing to winning, then the player's marginal contribution for that coalition is one; otherwise it is zero. A coalition $S$ is said to be a *swing* for player $i$ if $S$ is losing but $S \cup \{i\}$ is winning.

For player $i \in N$, let $\eta_i$ denote the number of swings, i.e.,:

$$\eta_i = \sum_{T_i} 1 \tag{3}$$

where $T_i$ is a losing coalition but $T_i \cup \{i\}$ is winning. The two versions of Banzhaf index are defined by expressing $\eta_i$ over different denominators.

**Definition 2.** *For player $i$, the absolute Banzhaf index ($\beta_i$) is defined as [2]:*

$$\beta_i = \eta_i / 2^{n-1} \tag{4}$$

**Definition 3.** *For player $i$, the normalized Banzhaf index ($\lambda_i$) is defined as [2]:*

$$\lambda_i = \eta_i / \Sigma_{i=1}^{n} \eta_i \tag{5}$$

Note that the normalized Banzhaf index sums to unity over the players: $\sum \lambda_i = 1$.

The problem of computing the Shapley value or the Banzhaf index for voting games is #P-complete [5, 12]. In order to overcome this problem, we present new approximation methods to find these indices.

# 3 Approximate Power Indices

The methods we propose are an extension of the one presented in [7]. In more detail, [7] is an approximation for the Shapley value for weighted voting games. Here we extend this to find approximates for the Shapley value and the Banzhaf index for both wieghted voting games and $k$-majority games. Section 3.1 deals with methods for weighted voting games and Section 3.2 with those for $k$-majority games.

## 3.1 Weighted Voting Game

The intuition behind the method proposed in [7] is as follows. As per Definition 1, in order to find a player's Shapley value, we first need to find his marginal contribution to all possible coalitions. For $n$ players, there are $2^{n-1}$ possible coalitions. Finding a player's marginal contribution to each of these $2^{n-1}$ possible coalitions is computationally infeasible. So instead of finding the marginal contribution to each possible coalition, this method finds a player's expected marginal contribution to random coalitions of size $X$ where $1 \leq X \leq n$. This is done by using the approximation rule $\mathcal{R}_1$ which is defined as follows.

Let the players' weights in $N$ be defined by any probability distribution function. Irrespective of the actual form of this function, let $\mu$ be the mean weight for the set of players and $\nu$ be the variance in the players' weights. From this set ($N$) if we randomly draw a sample, then the approximate sum of the players' weights in the sample is given by the following rule [8]:

$\mathcal{R}_1$: If $w_1, w_2, \ldots, w_X$ is a random sample of size $X$ drawn from *any* distribution with mean $\mu$ and variance $\nu$, then the sample sum has an *approximate Normal distribution*, $\mathcal{N}$, with mean $X\mu$ and variance $\frac{\nu}{X}$ (the larger the $n$ the better the approximation[2]).

We know, from Definition 1, that the Shapley value for a player is the expectation ($E$) of its marginal contribution to a coalition that is chosen randomly. The above rule is used to determine the Shapley value as follows.

[2] Also, for large $X$, any measurement done on a sample drawn with replacement is the same as that for a sample drawn without replacement [8].

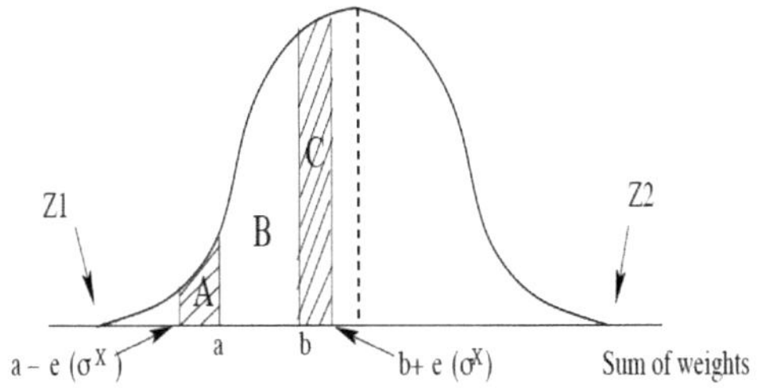


**Fig. 1.** A normal distribution for the sum of players' weights in a coalition of size $X$.

For player $i$ with weight $w_i$, let $\bar{\varphi}_i$ denote the approximate Shapley value. Also, let $X$ denote the size of a random sample drawn from $N$. The marginal contribution of player $i$ to this random sample is one if the total weight of the $X$ players in the sample is greater than or equal to $a = q - w_i$ but less than $b = q - \epsilon$ (where $\epsilon$ is an inifinitesimally small quantity). Otherwise, its marginal contribution is zero. Thus, the expected (approximate) marginal contribution of player $i$ (denoted $E\Delta_i^X$) to the sample coalition is the area under the curve defined by $\mathcal{N}(X\mu, \frac{\nu}{X})$ in the interval $[a, b]$. This area is shown as the region $B$ in Figure 1 (the dotted line in the figure is $X\mu$). Hence $i$'s approximate marginal contribution to $X$ is:

$$E\Delta_i^X = \frac{1}{\sqrt{(2\pi\nu/X)}} \int_a^b e^{-X\frac{(x-X\mu)^2}{2\nu}} dx. \tag{6}$$

And, as per Definition 1, $i$'s approximate Shapley value (denoted $\bar{\varphi}_i$) is the average of his expected marginal contribution to all possible coalitions:

$$\bar{\varphi}_i = \frac{1}{n} \sum_{X=1}^{n} E\Delta_i^X \tag{7}$$

The time complexity of this method is $\mathcal{O}(n)$ [7].

We now extend this method to find the Banzhaf index. For a game of $n$ players, let $T$ denote the number of possible coalitions of $X$ players, i.e., $T = C(n, X)$ is the number of combinations of $X$ items drawn from a set of $n$ items. Given this, player $i$'s total approximate marginal contribution to all coalitions of size $X$ is $C(n, X) \times E\Delta_i^X$ where $E\Delta_i^X$ is as computed in Equation 6. In other words, $i$'s approximate number of swings for coalitions of size $X$ is:

$$\bar{\eta}_i^X = C(n, X) \times E\Delta_i^X \tag{8}$$

Hence, $i$'s approximate number of swings to coalitions of all possible sizes ($1 \leq X \leq n$) is:

$$\bar{\eta}_i = \sum_{X=1}^{n} \bar{\eta}_i^X \tag{9}$$

As per Equation 4, $i$'s approximate absolute Banzhaf index ($\bar{\beta}_i$) is:

$$\bar{\beta}_i = \bar{\eta}_i/2^{n-1} \tag{10}$$

And as per Equation 5, $i$'s approximate normalised Banzhaf index ($\bar{\lambda}_i$) is:

$$\bar{\lambda}_i = \bar{\eta}_i/\delta \tag{11}$$

where $\delta = \sum_{i=1}^{n} \bar{\eta}_i$.

**Algorithm 1.** BanzhafIndexWVG($n$, $q$, $\mu$, $\nu$, $w$)

$N$: Set of players
$q$: Quota for the game
$\mu$: A $k$ element vector containing the mean weight of the players in $N$ for the $k$ weighted voting games
$\nu$: A $k$ element vector containing the variance in the weights of the players in $N$ for the $k$ weighted voting games
$w$: A vector of the player's weights

1: $T \Leftarrow 0$
2: **for** $i = 1$ to $n$ **do**
3: $\quad \bar{\eta}_i \Leftarrow 0$
4: $\quad$ **for** $X = 1$ to $n$ **do**
5: $\quad\quad a \Leftarrow q - w_i;\ b \Leftarrow q - \epsilon$
6: $\quad\quad E\Delta_i^X \Leftarrow \frac{1}{\sqrt{2\Pi\nu/X}} \int_a^b e^{-X\frac{(x-X\mu)^2}{2\nu}} dx$
7: $\quad\quad \bar{\eta}_i^X \Leftarrow E\Delta_i^X \times C(n, X)$
8: $\quad\quad \bar{\eta}_i \Leftarrow \bar{\eta}_i + \bar{\eta}_i^X$
9: $\quad$ **end for**
10: $\quad T \Leftarrow T + \bar{\eta}_i$
11: **end for**
12: **for** $i$=1 to $n$ **do**
13: $\quad \bar{\beta}_i \Leftarrow \eta_i/2^{n-1}$
14: $\quad \bar{\lambda}_i \Leftarrow \eta_i/T$
15: **end for**
16: **return** $\bar{\beta}$ and $\bar{\lambda}$

The above steps are described in Algorithm 1. In more detail, Step 1 does the initialization. In Step 2, we vary $X$ between 1 and $n$ and repeatedly do the following. Step 3 is another initialization. In Step 4, we repeatedly do the followig. We find player $i$'s approximate marginal contribution to the random coalition of size $X$. In Step 7, we use Equation 6 to find the approximate number of player $i$'s swings for coalitions of size $X$. In Step 8, we do the same for coalitions of all possible sizes. In Step 10, we find the approximate sum of the swings for all the $n$ players. Finally, in Step 13 (14), we find $i$'s approximate absolute (normalized) Banzhaf index.

**Theorem 1.** *The time to compute $\bar{\beta}_i$ is $\mathcal{O}(n^2)$, and that for $\bar{\lambda}_i$ is $\mathcal{O}(n^3)$.*

*Proof. Since the time to compute $X!$ is $\mathcal{O}(X)$ and the time to compute $E\Delta_i^X$ (as per Equation 6) is $\mathcal{O}(1)$, the time to compute $\bar{\eta}_i^X$ is $\mathcal{O}(n)$ (see Equation 8). From Equation 9 we get the time to compute $\bar{\eta}_i$ as $\mathcal{O}(n^2)$, and from Equation 10 we get the time to compute $\bar{\beta}_i$ as $\mathcal{O}(n^2)$. Note that $E\Delta_i^X$ depends on the weight of player i, so it is different for different players. However, $C(n, X)$ and $X!$ are the same for all the players. Thus, we need to find $C(n, X)$ and $X!$ just once and reuse these values to compute the Banzhaf index for all the players. So once we find $\bar{\beta}_i$ for a player $i$, the time to find $\bar{\eta}_j^X$ for $j \neq i$ is $\mathcal{O}(1)$. So the time to find $\bar{\eta}_j$ is $\mathcal{O}(n)$. Given this, the time to find $\bar{\beta}_j$ for all players such that $i \neq j$ is $\mathcal{O}(n^2)$. It follows that $\delta = \sum \bar{\eta}_i$ can be found in time $\mathcal{O}(n^3)$ and so can $\bar{\lambda}_i$.*

### 3.2 $k$-Majority Voting Game

We now extend the method described in [7] to $k$-majority games. The intuition behind the proposed method is as follows. As described in Section 2.2, a $k$-majority game is defined in terms of $k$ weighted voting games $v_j$ $(1 \leq j \leq k)$. Given this definition, we first find a player's approximate marginal contribution to $v_j$ $(1 \leq j \leq k)$ using the method in [7]. Then on the basis of these $k$ marginal contributions, we find an approximate marginal contribution for a $k$-majority game as follows.

For a random coalition $S_X$ of size $X$, the approximate marginal contribution of player $i$ to the game $v_1 \wedge \ldots \wedge v_k$ is 1 if the following conditions hold:

1. there is at least one game $v_j$ $(1 \leq j \leq k)$ for which $i$ is the swing player, and
2. for each game $v_j$, the value of $S_X \cup \{i\}$ is 1.

We first introduce some notation to formalise the above conditions and then find an approximate Shapley value. Let $S_X$ be a random sample (of size $X$) drawn from $N$. For game $v_t$ and player $i$, let $PL_i^t(S_X)$ (where $S_X \subset N - \{i\}$) denote the probability that the coalition $S_X$ is losing but $S_X \cup \{i\}$ is winning (i.e., for game $v_t$, the probability that the expected marginal contribution of $i$ to $S_X$ is 1). Also, for game $v_t$, let $PW^t(S_X)$ denote the probability that the coalition $S_X$ is winning (i.e., the probability that marginal contribution of $i$ to $S_X$ is 0). Finally, for game $v_t$, let $\mu^t$ denote the mean weight of the players, $\nu^t$ the variance in their weights, and $q^t$ the quota. Then, for a $k$-majority game, $i$'s expected marginal contribution to $S_X$ is:

$$kE\Delta_i^X = \sum_{j=1}^{k}\left(\prod_{f=1}^{j} PL_i^f(S_X) \times \prod_{g=j+1}^{k} PW^g(S_X)\right) \tag{12}$$

where $PL_i^t(S_X)$ is the area under the normal distribution $N(\mu^t, \nu^t)$ between the limits $q^t - w_i^t$ and $q^t - \epsilon$:

$$PL_i^t(S_X) = \frac{1}{\sqrt{(2\pi\nu^t/X)}} \int_{q^t - w_i^t}^{q^t - \epsilon} e^{-X\frac{(x - X\mu^t)^2}{2\nu^t}} dx \tag{13}$$

and $PW^t(S_X)$ is the area under the normal distribution $N(\mu^t, \nu^t)$ between the limits $q^t$ and $\infty$:

$$PW^t(S_X) = \frac{1}{\sqrt{(2\pi\nu^t/X)}}\left(\int_0^{q^t-w_i^w-1} e^{-X\frac{(x-X\mu^t)^2}{2\nu^t}}dx + \int_{q^t}^{\infty} e^{-X\frac{(x-X\mu^t)^2}{2\nu^t}}dx\right) \tag{14}$$

Given $PL_i^t(S_X)$ and $PW^t(S_X)$, the approximate Shapley value (as per Definition 1) for player $i$ for a $k$-majority game is:

$$\bar{\varphi}_i^k = \frac{1}{n}\sum_{X=1}^{n} kE\Delta_i^X \tag{15}$$

The above steps are described in Algorithm 2. We now present the time complexity of this method.

**Theorem 2.** *The time complexity of Algorithm 2 is $\mathcal{O}(k^2n)$.*

*Proof. The time to execute the for loop in Step 4 of Algorithm 2 is $\mathcal{O}(k^2)$. Since this for loop is within the for loop of Step 2 (which is executed $n$ times), the time complexity of Algorithm 2 is $\mathcal{O}(k^2n)$.*

We now extend the method described in Algorithm 1 (for Banzhaf index) to $k$-majority games. For player $i$', let $k\bar{\eta}_i^X$ denote the approximate number of swings for coalitions of size $X$. Then from Equation 8, we have the following:

$$k\bar{\eta}_i^X = C(n, X) \times kE\Delta_i^X \tag{16}$$

where $kE\Delta_i^X$ is as computed in Equation 12. Substituting Equation 12 in Equation 16 we get $k\bar{\eta}_i^X$. For player $i$, let $k\bar{\eta}_i$ be the approximate number of swings for coalitions of all possible sizes. Also, for player $i$, let $\bar{\beta}_i^k$ and $\bar{\lambda}_i^k$ denote the approximate absolute and normalised Banzhaf indices respectively. Then, player $i$'s approximate number of swings over coalitions of all possible sizes is:

$$k\bar{\eta}_i = \sum_{X=1}^{n} k\bar{\eta}_i^X \tag{17}$$

As per Equation 4, $i$'s approximate absolute Banzhaf index ($\bar{\beta}_i^k$) is:

$$\bar{\beta}_i^k = k\bar{\eta}_i/2^{n-1} \tag{18}$$

And as per Equation 5, $i$'s approximate normalised Banzhaf index ($\bar{\lambda}_i^k$) is:

$$\bar{\lambda}_i^k = k\bar{\eta}_i/\delta^k \tag{19}$$

where $\delta^k = \sum_{i=1}^{n} k\bar{\eta}_i$.

The above steps are detailed in Algorithm 3.

**Algorithm 2.** ShapleyValue-KMG($k, n, q, \mu, \nu, w$)

$k$: The number of weighted voting games
$n$: Number of players
$q$: Quota for the game
$\mu$: A $k$ element vector containing the mean weight of the players in $N$ for the $k$ weighted voting games
$\nu$: A $k$ element vector containing the variance in the weights of the players in $N$ for the $k$ weighted voting games
$w$: The player's weights for the $k$ weighted voting games

$T_i \Leftarrow 0$;
**for** $X = 1$ to $n$ **do**
  sum $\Leftarrow 0$
  **for** $j = 1$ to $k$ **do**
    prod $\Leftarrow 1$
    **for** $f = 1$ to $j$ **do**
      $a \Leftarrow q^f - w_i^f$; $b \Leftarrow q^f - \epsilon$
      prod $\Leftarrow$ prod $\times \frac{1}{\sqrt{2\Pi\nu^f/X}} \int_a^b e^{-X\frac{(x-X\mu^f)^2}{2\nu^f}} dx$
    **end for**
    **for** $g = j+1$ to $k$ **do**
      $a \Leftarrow q^g$; $b_1 \Leftarrow q^g - w_i^g - 1$; $b_2 \Leftarrow \infty$
      prod $\Leftarrow$ prod $\times \frac{1}{\sqrt{2\Pi\nu^g/X}} (\int_0^{b_1} e^{-X\frac{(x-X\mu^g)^2}{2\nu^g}} dx + \int_a^{b_2} e^{-X\frac{(x-X\mu^g)^2}{2\nu^g}} dx)$
    **end for**
    sum $\Leftarrow$ sum + prod
  **end for**
  $kE\Delta_i^X \Leftarrow$ sum
  $T_i \Leftarrow T_i + E\Delta_i^X$
**end for**
$\bar{\varphi}_i^k \Leftarrow T_i/n$
**return** $\bar{\varphi}_i^k$

**Theorem 3.** *The time to compute $\bar{\beta}_i^k$ is $\mathcal{O}(k^2n^2)$ and that for $\bar{\lambda}_i^k$ is $\mathcal{O}(n^3k^2)$.*

*Proof.* *As per Equation 12, the time to find $kE\Delta_i^X$ is $\mathcal{O}(k^2)$. Also, as per Equation 16, the time to find $k\bar{\eta}_i^X$ is $\mathcal{O}(nk^2)$. From Equation 17, we get the time to find $k\bar{\eta}_i$ as $\mathcal{O}(n^2k^2)$. From Equation 18, we know that the time find $\bar{\beta}_i^k$ the same as the time to find $k\bar{\eta}_i$. Given this, the time to compute $\delta^k$ is $n$ times the time to compute $k\bar{\eta}_i$. Hence, from Equation 19, we get the time to compute $\bar{\lambda}_i^k$ as $\mathcal{O}(n^3k^2)$.*

Now, the quality of an approximation method is evaluated on the basis of its running time and also its approximation error. To this end, the following section conducts error analysis for the proposed methods.

**Algorithm 3.** BanzhafIndex-KMG($k$, $n$, $q$, $\mu$, $\nu$, $w$)

$k$: The number of weighted voting games
$n$: Number of players
$q$: Quota for the game
$\mu$: A $k$ element vector containing the mean weight of the players in $N$ for the $k$ weighted voting games
$\nu$: A $k$ element vector containing the variance in the weights of the players in $N$ for the $k$ weighted voting games
$w$: An array of player's weights for the $k$ weighted voting games

$T_i \Leftarrow 0$;
**for** $i = 1$ to $n$ **do**
  **for** $X = 1$ to $n$ **do**
    sum $\Leftarrow 0$
    **for** $j = 1$ to $k$ **do**
      prod $\Leftarrow 1$
      **for** $f = 1$ to $j$ **do**
        $a \Leftarrow q^f - w_i^f; b \Leftarrow q^f - \epsilon$
        prod $\Leftarrow$ prod $\times \frac{1}{\sqrt{2\Pi\nu^f/X}} \int_a^b e^{-X\frac{(x-X\mu^f)^2}{2\nu^f}} dx$
      **end for**
      **for** $g = j + 1$ to $k$ **do**
        $a \Leftarrow q^g; b_1 \Leftarrow q^g - w_i^g - 1; b_2 \Leftarrow \infty$
        prod $\Leftarrow$ prod $\times \frac{1}{\sqrt{2\Pi\nu^g/X}} \left( \int_0^{b_1} e^{-X\frac{(x-X\mu^g)^2}{2\nu^g}} dx + \int_a^{b_2} e^{-X\frac{(x-X\mu^g)^2}{2\nu^g}} dx \right)$
      **end for**
      sum $\Leftarrow$ sum + prod
    **end for**
    $kE\Delta_i^X \Leftarrow$ sum
    $k\bar{\eta}_i^X \Leftarrow kE\Delta_i^X \times C(n, X)$
    $k\bar{\eta}_i \Leftarrow k\bar{\eta}_i + k\bar{\eta}_i^X$
  **end for**
  $T \Leftarrow T + k\bar{\eta}_i$
**end for**
**for** $i$=1 to $n$ **do**
  $\bar{\beta}_i^k \Leftarrow \eta_i / 2^{n-1}$
  $\bar{\lambda}_i^k \Leftarrow \eta_i / T$
**end for**
**return** $\bar{\beta}^k$ and $\bar{\lambda}^k$

## 4 Error Analysis

We first formalize the idea of *error* and then derive the formula for measuring the error in the approximate power indices of Section 3. The concept of error relates to a measurement made of a quantity which has an exact value [18, 3]. Obviously, it cannot be determined exactly how far off a measurement is from the exact value; if this could be done, it would be possible to just give the more accurate, corrected value. Thus, error has to do with *uncertainty* in measurements that nothing can be done about. However, although it is not possible to do anything about such an error, it can be characterized in terms of two essential components [18, 3]:

1. a numerical value giving the best "estimate" possible of the quantity measured, and
2. the degree of *uncertainty* associated with this estimated value.

For example, if the estimate of a quantity is $x$ and the uncertainty is $e(x)$ the quantity would lie in $x \pm e(x)$. For sampling based methods, uncertainty is characterized in terms of *standard error* [18] which is analogous to the algorithmic term *absolute error*. This error is equal to the absolute difference between the approximate and its exact counterpart [1]. We first find this error for our approximate Shapley value and then compare it with the error for our approximate Banzhaf index.

### 4.1 Absolute Error

Standard error, which we use to measure the absolute error, is defined as follows [18, 3]:

**Definition 4.** *Standard error is defined as the standard deviation for a set of measurements divided by the square root of the number of measurements.*

Given this definition, for a weighted voting game, let $e(\sigma^X)$ be the absolute error in the approximate sum of weights for a random coalition of size $X$ where:

$$\begin{aligned} e(\sigma^X) &= \sqrt(\nu/X)/\sqrt(X) \\ &= \sqrt(\nu)/X. \end{aligned} \tag{20}$$

Then let $e(E\Delta_i^X)$ denote the error in the approximate marginal contribution for player $i$ (given in Equation 6). This error is obtained by propagating the error in Equation 20 to the error in a player's expected marginal contribution given in Equation 6. In Equation 6, $a$ and $b$ are the lower and upper limits for the sum of the players' weights for a coalition of size $X$. Since the error in this sum is $e(\sigma^X)$, the actual values of $a$ and $b$ lie in the intervals $a \pm e(\sigma^X)$ and $b \pm e(\sigma^X)$ respectively. Hence, the error in Equation 6 is either the probability that the sum lies between the limits $a - e(\sigma^X)$ and $a$ (i.e., the area under the curve defined by $\mathcal{N}(X\mu, \frac{\nu}{X})$ between $a - e(\sigma^X)$ and $a$, which is the shaded region $A$ in Figure 1) or the probability that the sum of weights lies between the limits $b$ and $b + e(\sigma^X)$ (i.e., the area under the curve defined by $\mathcal{N}(X\mu, \frac{\nu}{X})$ between $b$ and $b + e(\sigma^X)$, which is the shaded region $C$ in Figure 1). More specifically, the error is at most the maximum of these two probabilities:

$$\begin{aligned} e(E\Delta_i^X) = \frac{1}{\sqrt(2\pi\nu/X)} \times \text{MAX}\Bigg( & \int_{a-e(\sigma^X)}^{a} e^{-X\frac{(x-X\mu)^2}{2\nu}} dx, \\ & \int_{b}^{b+e(\sigma^X)} e^{-X\frac{(x-X\mu)^2}{2\nu}} dx \Bigg) \end{aligned} \tag{21}$$

On the basis of the above error, we find the error in the Shapley value by using the following standard error propagation rules. Let $x$ and $y$ be two random variables with errors $e(x)$ and $e(y)$ respectively. Then, from [18] we have the following propagation rules:

R1 The error in the random variable $z = x + y$ is:

$$e(z) = e(x) + e(y)$$

R2 If $z = kx$ where the constant $k$ has no error, then the error in $z$ is:

$$e(z) = |k|e(x)$$

R3 The error in the random variable $z = x \times y$ is:

$$e(z) = e(x) + e(y)$$

### 4.2 Absolute Error for Weighted Voting Games

Using the above rules, the error in the Shapley value (given in Equation 7) is obtained by propagating the error in Equation 21 to all coalitions between the sizes $X = 1$ and $X = n$. This error (denoted $e(\bar{\varphi}_i)$) is:

$$e(\bar{\varphi}_i) = \frac{1}{n} \sum_{X=1}^{n} e(E\Delta_i^X) \tag{22}$$

Note that we are finding the absolute error for the Shapley value. Here, it is interesting to note that a related concept for characterising the quality of approximation is *performance ratio*. Roughly speaking, this is the ratio of an approximate solution and its exact counterpart [1]. The problem of approximating the Shapley value such that the approximation ratio is bounded by a constant is intractable unless P=NP [6]. In future, it would be interesting to obtain a similar result for the absolute error as well.

We now turn to the error in the approximate Banzhaf index. Using the error propagation rules (R1, R2, and R3), we get the error in the $\bar{\eta}_i^X$ (see Equation 8) as:

$$e(\bar{\eta}_i^X) = e(E\Delta_i^X) \times C(n, X). \tag{23}$$

Given Equation 23, the error in $\bar{\eta}_i$ (see Equation 9) is:

$$e(\bar{\eta}_i) = \sum_{X=1}^{n} e(\bar{\eta}_i^X) \tag{24}$$

From Equations 24 and 10, we get the error in $\bar{\beta}_i$ as:

$$e(\bar{\beta}_i) = e(\bar{\eta}_i)/2^{n-1} \tag{25}$$

And, as per Equaton 11, the error in $\bar{\lambda}_i$ is:

$$e(\bar{\lambda}_i) = e(\bar{\eta}_i) + \sum_{i=1}^{n} e(\bar{\eta}_i) \tag{26}$$

The above equations lead to the following observation for our methods:

**Observation.** For a given weighted voting game, we have the following relationship: the approximation error in a player's normalized Banzhaf index is higher than the error in its absolute Banzhaf index and the error in its Shapley value $\big(e(\bar{\lambda}_i) > e(\bar{\varphi}_i)$ and $e(\bar{\lambda}_i) > e(\bar{\beta}_i)\big)$.

### 4.3 Error for $k$-Majority Games

On the basis of the results of Section 4.2, we now analyze the error for $k$-majority games. Before doing so, we introduce some notation. Let $e(\sigma_t^X)$ be the error in the approximate sum of weights of $S_X$ for game $t$. Let $e(PL_i^t(S_X))$ and $e(PW^t(S_X))$ denote the errors in $PL_i^t(S_X)$ and $PW^t(S_X)$ respectively. These two errors are obtained in the same way as we obtained $e(E\Delta_i^X)$ in Equation 21. Hence we have:

$$e(PL_i^t(S_X)) = \frac{1}{\sqrt{(2\pi\nu^t/X)}} \times \mathrm{MAX}\Bigg(\int_{q^t-\epsilon}^{q^t-\epsilon+e(\sigma_t^X)} e^{-X\frac{(x-X\mu^t)^2}{2\nu^t}}\,dx, \int_{q^t-w_i^t-e(\sigma_t^X)}^{q^t-w_i^t} e^{-X\frac{(x-X\mu^t)^2}{2\nu^t}}\,dx\Bigg) \tag{27}$$

and

$$e(PW^t(S_X)) = \frac{1}{\sqrt{(2\pi\nu^t/X)}} \times \Bigg(\int_{q^t-w^t-1}^{q^t-w_i^t-1-e(\sigma_t^X)} e^{-X\frac{(x-X\mu^t)^2}{2\nu^t}}\,dx + \int_{q^t-e(\sigma_t^X)}^{q^t} e^{-X\frac{(x-X\mu^t)^2}{2\nu^t}}\,dx\Bigg). \tag{28}$$

For $k$-majority games, let $e(kE\Delta_i^X)$ denote the error in $i$'s marginal contribution to a random coalition of size $X$, and let $e(\bar{\varphi}_i^k)$ denote the error in $i$'s Shapley value. From rule R3, we get:

$$e(kE\Delta_i^X) = \sum_{j=1}^{k}\Bigg(\sum_{f=1}^{j} e(PL_i^f(S_X)) + \sum_{g=j+1}^{k} e(PW^g(S_X))\Bigg) \tag{29}$$

So the error in $i$'s Shapley value is:

$$e(\bar{\varphi}_i^k) = \frac{1}{n}\sum_{X=1}^{n} e(kE\Delta_i^X) \tag{30}$$

We now analyze the error in Banzhaf index. Combining Equation 29 with Equation 16, we get the error in $k\bar{\eta}_i^X$ as:

$$e(k\bar{\eta}_i^X) = e(kE\Delta_i^X) \times C(n, X) \tag{31}$$

The error in $k\bar{\eta}_i$ (see Equation 17) is:

$$e(k\bar{\eta}_i) = \sum_{X=1}^{n} e(k\bar{\eta}_i^X) \tag{32}$$

The error in $\bar{\beta}_i^k$ (see Equation 18) is:

$$e(\bar{\beta}_i^k) = e(k\bar{\eta}_i)/2^{n-1} \tag{33}$$

The error in $\bar{\lambda}_i^k$ (see Equation 19) is:

$$e(\bar{\lambda}_i^k) = e(k\bar{\eta}_i) + \sum_{i=1}^{n} e(k\bar{\eta}_i). \tag{34}$$

From the above equations, we make the following observation regarding our methods.

**Observation.** For a given $k$-majority game, we have the following relationship: the approximation error in a player's normalized Banzhaf index is higher than the error in its absolute Banzhaf index and the error in its Shapley value $\left(e(\bar{\lambda}_i^k) > e(\bar{\varphi}_i^k)\right.$ and $\left.e(\bar{\lambda}_i^k) > e(\bar{\beta}_i^k)\right)$.

## 5 Related Work

A number of approximation methods have been proposed for finding an approximate Shapley value. These include [9, 11, 19]. The method proposed in [9] is based on computing an approximate Shapley value by making measurements on random samples of coalitions. However the method does not specify how the samples need to be drawn. It is important to know how to draw samples because this is a key factor that determines the quality of approximation. In contrast, our method is based on the approximation rule defined in Section 3.1 and does not require making measurements on random samples. The method proposed in [11] uses a different randomization method from ours but like our method, it too does not require drawing random samples. Finally, [19] presented a randomization method for an approximate Shapley value in the context of task oriented domains. Also, like our method, [11, 19] have linear time complexity. Hence, in future, we need to compare the approximation error for these two methods with that for ours.

## 6 Conclusions and Future Work

The Shapley value and Banzhaf index are two well known indices for measuring a the power a player has in a voting game. However, the problem of computing these indices is computationally hard. To overcome this problem, we presented new approximation methods for computing these indices. Although the proposed methods have polynomial time complexity, finding an approximate Shapley value using them is easier than finding an approximate Banzhaf index. We also found the *absolute error* for our methods and showed that this error for the Shapley value is lower than that for the Banzhaf index. In future, we need to find the bounds on these errors.

## References


1. Ausiello, G., Crescenzi, P., Gambosi, G., Kann, V., Marchetti-Spaccamela, A., Protasi, M.: Complexity and approximation: Combinatorial optimization problems and their approximability properties. Springer, Heidelberg (2003)
2. Banzhaf, J.F.: Weighted voting doesn't work: A mathematical analysis. Rutgers Law Review 19, 317–343 (1965)
3. Bork, P.V., Grote, H., Notz, D., Regler, M.: Data Analysis Techniques in High Energy Physics Experiments. Cambridge University Press, Cambridge (1993)
4. Conitzer, V., Sandholm, T.: Complexity of constructing solutions in the core based on synergies among coalitions. Artificial Intelligence Journal 170, 607–619 (2006)
5. Deng, X., Papadimitriou, C.H.: On the complexity of cooperative solution concepts. Mathematics of Operations Research 19(2), 257–266 (1994)
6. Elkind, E., Goldberg, L.A., Goldberg, P., Wooldridge, M.: Computational complexity of weighted threshold games. In: Proceedings of the National Conference on Artificial Intelligence (AAAI 2007) (2007)
7. Fatima, S.S., Wooldridge, M., Jennings, N.R.: A randomized method for the Shapley value for the voting game. In: Proceedings of the Sixth International Conference on Autonomous Agents and Multi-Agent Systems, pp. 955–962 (2007)
8. Francis, A.: Advanced Level Statistics. Stanley Thornes Publishers (1979)
9. Mann, I., Shapley, L.S.: Values for large games iv: Evaluating the electoral college by monte carlo techniques. Technical report, The RAND Corporation, Santa Monica (1960)
10. Osborne, M.J., Rubinstein, A.: A Course in Game Theory. The MIT Press, Cambridge (1994)
11. Owen, G.: Multilinear extensions of games. Management Science 18(5), 64–79 (1972)
12. Prasad, K., Kelly, J.S.: NP-completeness of some problems concerning voting games. International Journal of Game Theory 19, 1–9 (1990)
13. Rahwan, T., Jennings, N.R.: An algorithm for distributing coalitional value calculations among cooperating agents. Artificial Intelligence Journal 171, 535–567 (2007)
14. Sandholm, T., Lesser, V.: Coalitions among computationally bounded agents. Artificial Intelligence Journal 94(1), 99–137 (1997)
15. Shapley, L.S.: A value for n person games. In: Roth, A.E. (ed.) The Shapley value, pp. 31–40. University of Cambridge Press, Cambridge (1988)
16. Shehory, O., Kraus, S.: A kernel-oriented model for coalition-formation in general environments: Implemetation and results. In: Proceedings of the National Conference on Artificial Intelligence (AAAI 1996), pp. 131–140 (1996)
17. Shehory, O., Kraus, S.: Methods for task allocation via agent coalition formation. Artificial Intelligence Journal 101(2), 165–200 (1998)
18. Taylor, J.R.: An introduction to error analysis: The study of uncertainties in physical measurements. University Science Books (1982)
19. Zlotkin, G., Rosenschein, J.: Coalition, cryptography, and stability: mechanisms foe coalition formation in task oriented domains. In: Proceedings of the National Conference on Artificial Intelligence (AAAI 1994), pp. 432–437 (1994)

# Author Index

GPSR Compliance

*The European Union's (EU) General Product Safety Regulation (GPSR) is a set of rules that requires consumer products to be safe and our obligations to ensure this.*

*If you have any concerns about our products, you can contact us on ProductSafety@springernature.com*

In case Publisher is established outside the EU, the EU authorized representative is:

Springer Nature Customer Service Center GmbH
Europaplatz 3
69115 Heidelberg, Germany

Batch number: 10614746

Printed by Printforce, the Netherlands